"Ben Austen has emerged, over the last five years, as one of the most serious and thoughtful new American reporters. This book was years in the making and, in some way, Austen's whole life in the making. In it a neighborhood becomes a character, a protagonist, but the character has inside it real human beings. Austen convinced me that my understanding of what goes on inside 'the projects' had been about as deep as a cop show. We need more books like this from him."

—John Jeremiah Sullivan, author of *Pulphead*

"One of the things I love about this book is that Austen has masterfully woven together these deeply intimate stories of the residents at Cabrini against the backdrop of critical public policy decisions. Ultimately this book is about how we as a country acknowledge and deal with the very poor."

—Alex Kotlowitz, author of *There Are No Children Here*

"It's a sad, infuriating, complicated tale that has been told bit by bit over the decades but never assembled into such a comprehensive, readable book." —Mary Schmich, *Chicago Tribune*

"A finely crafted biography of an urban community."

—*Library Journal*

"A tenant's-eye view of life in one of the most infamous public housing projects . . . of a community fighting to live in a system stacked against them." —*CityLab*

"We can now better consider remedies by contemplating the lessons *High-Risers* offers."

—Richard Rothstein, author of *The Color of Law*

"Austen demonstrates the centrality of Cabrini-Green to Chicago's sense of itself." —*Chicago Reader*

"A weighty and robust history of a people disappeared from their own community." —*Kirkus Reviews*

"Austen lends a novelistic eye to moments large and small. . . . Austen's book resists gentrification's whitewashing effect by painting a vibrant portrayal of the communities, individual lives and politics that intersected at Cabrini-Green before the wrecking ball swung." —*Newcity Lit*

"With this book, Austen joins Natalie Y. Moore, Eve Ewing, and Alex Kotlowitz as one of Chicago's great chroniclers of the devastating effects of bad public policy. . . . Should be required reading." —*Chicago Review of Books*

"Urban planners in particular will find this an instructive guide, or, perhaps more importantly, a cautionary tale about a failed attempt to provide affordable housing for the poor." —*Publishers Weekly*

"[Austen] ties the history of Cabrini-Green to broad economic, political, and social trends that played a pivotal role in the creation and undoing not only of Cabrini-Green, but also of much of America's public housing. . . . The high-rises are gone and, in time, the high-risers will go with them. But the lessons of Cabrini-Green still weigh on us all." —*South Side Weekly*

HIGH-RISERS

CABRINI-GREEN AND

THE FATE OF AMERICAN

PUBLIC HOUSING

■

BEN AUSTEN

HARPER

NEW YORK · LONDON · TORONTO · SYDNEY

HARPER

HIGH-RISERS. Copyright © 2018 by Ben Austen. All rights reserved. Printed in
the United States of America. No part of this book may be used or reproduced
in any manner whatsoever without written permission except in the case of
brief quotations embodied in critical articles and reviews. For information,
address HarperCollins Publishers, 195 Broadway, New York, NY 10007.

HarperCollins books may be purchased for educational, business, or
sales promotional use. For information, please email the Special Markets
Department at SPsales@harpercollins.com.

FIRST HARPER PAPERBACKS EDITION PUBLISHED 2019.

Designed by Fritz Metsch

Library of Congress Cataloging-in-Publication Data has been applied for.

ISBN 978-0-06-223507-7 (pbk.)

23 24 25 26 27 LBC 8 7 6 5 4

For my family and my city

'CABRINI·GREEN'   AS IT WAS....

GOLD COAST

LA SALLE

EL

SEWARD PARK

JENNER SCH.

DURSO PARK

HUDSON

CHICAGO

LARRABEE

CABRINI ROWHOUSES

MONTGOMERY WARD WAREHOUSE

KINGSBURY

CHICAGO RIVER

HALSTED

C. ROBERT GORDON JUNE 2017

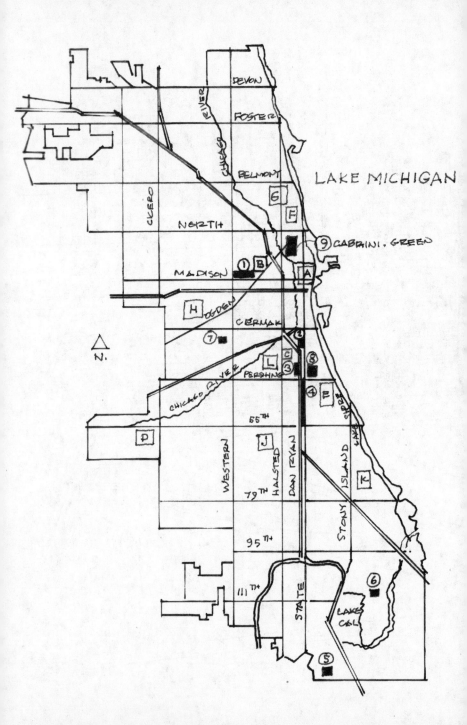

KEY

1. HENRY HORNER HOMES
2. DEARBORN HOMES
3. WENTWORTH GARDENS
4. ROBERT TAYLOR HOMES
5. ALTGELD GARDENS
6. TRUMBULL PARK HOMES
7. LAWNDALE GARDENS
8. IDA B. WELLS HOMES
9. CABRINI-GREEN

A. LOOP
B. BULLS STADIUM
C. WHITE SOX PARK
D. MIDWAY AIRPORT
E. BRONZEVILLE
F. GOLD COAST
G. LINCOLN PARK
H. NORTH LAWNDALE
J. ENGLEWOOD
K. SOUTH SHORE
L. BRIDGEPORT

© ROBERT GORDON

Moving through the rubble and babble down the locked-up, abandoned, ravished, sin-sacked and mildewed city; turning the cane about as if it were a symbolic key to the city given a visitor; a holy witness from another country come to save even you, and especially you. Nathaniel brooded now as they passed the government projects—adjacent to Rachel's house—and the fad-ridden negro youths, harmonizing in a doorway, who apparently didn't know that their home was far over Jordan.

—LEON FORREST, *The Bloodworth Orphans*

It's not just buildings.
It's not a place,
It's a feeling.
Since we all confess,
To be raised in Cabrini was a blessing. . . .
Cabrini is down but not out.
Have no doubt, Cabrini is God's goods stretched out.

—MICHAEL MCCLARIN, Cabrini-Green resident

Put the city up; tear the city down;
    put it up again; let us find a city.

—CARL SANDBURG, "The Windy City"

Contents

PART ONE

A HOME OVER JORDAN

1

Portrait of a Chicago Slum

TUCKED INTO THE elbow where the river tacks north, just beyond the Loop and a mile from Lake Michigan, it is as historic a neighborhood as there is in Chicago. In 2016, it was named one of the city's best places to live. A couple of generations earlier, and more than a century after the banks of the Near North Side were settled, surveyors from the Chicago Housing Authority walked its narrow streets, confirming with every step their belief that it was a slum beyond salvation. The field team from the CHA dodged trucks and trash heaps, careful lest they plunge into the open trenches dug for coal in front of the dwellings. The year was 1950, the quickening after the war, in the nation's "second city," yet everything looked to be of the benighted past. Almost all the buildings dated to the previous century. Many of them were cheap frame constructions slapped up after the Great Fire of 1871, temporary emergency shelters turned permanent. In their notebooks, the surveyors tallied the area's deprivations: nearly half of the 2,325 homes were without a bath or shower, many had no private toilet, and all but a few relied on coal stoves for heat. Over the previous decade, the population in the twenty-five square blocks had swelled to 3,600 families, increasing by 50 percent, yet only a single new residential building had been added. Flimsy partitions carved up the apartments into multiple units. "Excuse the appearance of this place," a housewife apologized as she welcomed the researchers into her subdivided home. "But we hardly have room to put ourselves someplace and there just ain't room for anything else." Despite the conditions, rents had jumped by 70 percent. Landlords overcharged for their firetraps.

The following year, the CHA issued its report, *Cabrini Extension Area: Portrait of a Chicago Slum*, which depicted in lurid detail the neighborhood the agency hoped to replace. "Houses, black with age and weathered with soot, lean precariously, and their uneven roof lines form crazy-quilt patterns against the sky. Chimneys tilt, eaves sag, rags stick out from broken windows, and doors without knobs stand open. There are few backyards. There can't be, when most of the lots contain two houses." Even the cover page sought to convey the ghetto in high squalor: a trompe l'oeil effect made the paper look burned and crumpled, as if found in one of the grubby alleys; the title was lettered in thick-markered graffiti script, with a drawing of a cockroach scuttling past the second "i" in "Cabrini."

The employees of the Chicago Housing Authority in 1950 weren't paper-pushing functionaries; they were self-proclaimed liberal do-gooders, many of them coming to the agency from social work. Their portrait of the Near North Side was meant to offend: they believed the slums of Chicago were killing people. House fires, infant mortality, pneumonia, tuberculosis, all occurred there at many times the rate found in the rest of the city. Poor housing conditions, the CHA noted, were contributing as well to high incidences of divorce, juvenile delinquency, and crime. The staff saw its work as a rescue mission: they needed to rid the city of blight. "Houses work magic," their boss at the agency, Elizabeth Wood, would say. "Give these people decent housing and the better forces inside them have a chance to work. Ninety-nine percent will respond."

Wood was an unlikely government official in the Chicago of the Democratic machine. She'd previously taught poetry at Vassar College and published a novel about an unhappily married woman who imparts her frustrations onto her children ("A psychological study of merciless persecution," a reviewer wrote). She moved to Chicago to work for a welfare agency but found the job ineffectual; she wanted to do more than scribble notes as desperate clients detailed their wants. When the CHA was formed in 1937, she took over as its executive director. Private enterprise had failed to provide the agency's minimal requirement of a "decent, safe and sanitary" home for all. The Near North Side district was just one of the woeful examples that the CHA brandished as proof. Wood feared not that the city's new public housing projects might be too large, coming to define an area as low-rent, but that they wouldn't be large enough to counteract the ravages of poverty and disrepair around them. "If it is not bold," she said, "the result will be a series of small projects, islands in a wilderness of slums beaten down by smoke, noise and fumes."

One of the islands Wood and her team hoped to expand was next to the Near North Side slum. In 1942, the CHA opened the Frances Cabrini Homes, 586 dwellings in barracks-style two- and three-story buildings. Federal rules established that public housing be built to minimum standards, using materials and designs unmistakably inferior to those found in market-rate housing. The Cabrini rowhouses were simple and unadorned, arranged in parallel columns like lines of parked tractor trailers. But in the neglected river district, they stood out as an oasis of order and modernity. "Like a challenge to the existing decay," the CHA declared. Each of the Cabrini Homes featured a gas stove, an electric refrigerator, a private bath, and its own heat controls. The buildings were made of "fireproof brick." And the development was laid out so that parents could watch from their apartments as children played in communal courtyards. "When you come upon one of Chicago's public housing developments, it is like stepping into a different world," the CHA rhapsodized in an early brochure. "Everywhere you see green— green of lawns, green of shrubbery, green of trees. Pleasant, vine-covered buildings stand in harmonious groups, with plenty of space left for sun and air and children's play. Everywhere you see gardens, and overhead stretches a sky that somehow looks bluer and sunnier than it did in the slums."

The Near North Side was largely Italian for much of the first half of the twentieth century. But a small black settlement formed there as well. The federal government restricted overseas immigration after the First World War, and so the factories along the river had vacancies they needed to fill. Most people also made great efforts to live far from the area's polluted worksites and ramshackle homes. Thus, African Americans were able to move in.

That was an anomaly in segregated Chicago. As African Americans turned their backs on the South, their population in Chicago more than doubled between 1910 and 1920, from 44,000 to 109,000, and then more than doubled again over the next decade and again over the next twenty years, reaching half a million by 1950. Up until the 1940s, almost all of these newcomers moved to the South Side, in what was called the "Black Belt," a broadening strip of land that extended south from the downtown business district.

The Europeans who made a home along the Chicago River's North Fork were free to test the private market elsewhere, even if affordable options were scarce. During the forties, the vacancy rate in the city fell to less than 1 percent, a total of eight thousand available units for the entirety of Chicago. But African Americans were forced to contend with a wholly separate real estate system. White neighborhoods established racial covenants, bylaws that barred homeowners from selling to African Americans. At one point, 85 percent of Chicago was covered under these restrictions. Even after the US Supreme Court outlawed the practice, in 1948, enforcement and legal recourse were negligible, and neighborhoods found less subtle means, such as assaults and firebombings, at least as effective. The federal government deemed existing black neighborhoods too risky for insured mortgages, coloring these areas red on its maps. "Redlining" meant African Americans could rarely purchase property in their own communities, except through predatory rent-to-own contract sales, in which buyers made inflated monthly payments but amassed no equity in the property. If or when they were evicted before the final payment, for any number of infractions, they lost the home, the down payment, and all the preceding monthly installments. After escaping the caprices of the Jim Crow South, drawn to Chicago by visions of a "promised land," African Americans found themselves at the mercy of a speculative housing system in the North unjust and unpredictable in its own rights.

Landlords in black neighborhoods enjoyed both overwhelming demand for their properties and a captive market. They not only charged high rents for their run-down dwellings, but also divided existing apartments into numerous "kitchenette" units. The practice, while common throughout overcrowded Chicago, was epidemic in the Black Belt. Within the same square footage, the number of occupants—and the amount of revenue—increased exponentially. Cut a six-family walk-up in half to house twelve families, into separate one-room apartments to make many more. There was no economic incentive for landlords to fix up their South Side properties; redlining meant that banks wouldn't loan them money for the work anyway.

With too many families crammed into airless wood-frame dwellings, forced to use alternative heating and cooking methods, with exposed wires and extension cords snaking in every direction from improvised walls and transoms to plug into the one or two overloaded circuits, fires were rampant. And because the kitchenettes were divided by nailed-up doorways and partitions that were themselves flammable, and because they lacked windows and safe exits, the fires too often proved deadly. "The kitchenette is our prison, our death sentence without a trial, the new form of mob violence that assaults not only the lone individual, but all of us in its ceaseless attacks," Richard Wright, who came to Chicago in 1927 from Mississippi by way of Memphis, lamented in his *12 Million Black Voices*. "The kitchenette is the funnel through which our pulverized lives flow to ruin and death on the city pavement, at a profit."

In the years following the Second World War, some twenty-three square miles of Chicago was said to be blighted, a tenth of the entire city; a quarter million homes, one-fourth of Chicago's total, were considered substandard. Chicagoans of all backgrounds were in need of the assistance of government-run public housing. But in the black slums of the South Side the need was greatest. "Due to his disadvantageous position in the present housing market, the Negro is the chief victim of excessive rents," the *Cabrini Extension Area* report concluded.

DOLORES WILSON

ONE OF WRIGHT'S twelve million, a woman in her twenties who felt condemned to her South Side tenement, was a lifelong Chicagoan named Dolores Wilson. At the start of the 1950s, she and her husband, Hubert, were parents to five children, ages eight to one. "Five snotty-nosed kids running around," Dolores liked to say with feigned annoyance. The Wilsons had a one-room basement apartment on the 6000 block of South Prairie Avenue, beside the grinding rattle of the El train and the trolley on Sixty-First Street. The children slept on one side of the room, on a pullout couch, and she and Hubert in a bed along the opposite wall. "With that arrangement," Dolores would say, "I'm not sure how the last child ever got made." The shower was propped up in the kitchen. The one window opened onto an alley. To use the toilet, they had to walk out their door and down the hallway, then past the laundry room to the bathroom the Wilsons shared with a family renting the back side of the partitioned basement.

She never felt safe in that building. One night while Hubert was at work, a man tried to break in through the window. "Oh, Lord," Dolores cried while watching the guy struggle to shimmy his way inside, the children asleep not six feet away. She managed to take out a little pistol Hubert had left her and call the police, the phone trembling in one hand and the gun in the other. An officer who answered said she could go ahead and shoot the burglar but only after he set foot in the apartment. Luckily, the man noticed her and took off.

Dolores wasn't one to complain—or rather, she had a way of doing it drolly, a genial-sounding protest, continuing all the while to make the best of a situation. She bought material from a five-and-dime and cut a red canopy for the mirror and the apartment's sole window; she found red dishes and a red-checkered tablecloth to match, decorating the apartment the way she liked it. "It doesn't matter where you are," she liked to say. "It's after you put your fingerprint on it. Then it's your home." Dolores Wilson had such a pleasant-seeming brightness about her, her voice soft and high like a confection, that she'd laugh out of frustration or spite and then feel the need to explain that it wasn't intended to be a funny laugh. She called most people "Dear," though not infrequently she meant it wryly. "I try to get along with everybody, even the ones I don't get along with."

She was careful to say, "Thank you, Father, for a roof over our heads." Everyone, at least, needed that. But it could feel like too much to cope with when the cold cut through the walls, or the structure meant to house your family might be killing them. The *Chicago Defender*, the city's leading black newspaper, kept count of the casualties from South Side house fires. "Negro children and women are dying like rats in fires in dilapidated homes unfit for human habitation, homes that are in reality firetraps which should have been condemned long ago by responsible officials," the paper wrote in one of many reports. Dolores often heard the sirens of the fire trucks. She knew that if her tenement went up in flames, they likely wouldn't make it out of the basement alive.

The Wilsons had moved a few times within the confines of their neighborhood, but the other apartments were no better. Options for them were limited. Landlords told her she had too many children, or that children older than toddlers caused trouble. Dolores and Hubert paid $10 to a real estate agent who promised to provide them with a special list of quality apartments. Dolores traveled to each apartment the agent gave her, finding herself in front of yet another South Side firetrap. She'd double-check that the address matched what she'd written down in her carefully looping script. The six-flats before her listed to one side, with rotting wood or missing bricks. Inside, the floors drooped, the walls buckled, and the ceilings leaked. The plaster and paint crumbled about could poison her children. "Uh-uh," she'd say, backing away, as the smells of greens and other food cooking from a dozen kitchenettes assailed her. "It made me want to go in there and fix a plate, but I didn't want to move in," she'd say. "If their food is loud, you know all the noise they're going to make." Ten dollars was a fortune to them, but after several of these trips Dolores had to accept that they'd been had.

Born Dolores Zanders, in 1929, at Chicago's Cook County Hospital, she grew up on that same block of 6000 South Prairie. Her mother's mother was an adventurer, a woman of means, who gave birth to each of her four children in a different state. Dolores's mother was originally from eastern Ohio, coal country, and she moved with her family to Chicago. Dolores's father followed a brother up north from Georgia. Her parents met in high school, and they settled into the apartment on Prairie as the block's last remaining Jewish family was set to depart. Dolores was one of their five children, and they lived well there. Dolores's mother worked as an assistant precinct captain for the local Democratic machine boss. Her father had a job as a presser and a tailor, even during the Depression years. He kept his pants pleated, his shoes polished to a high shine, and a satin-banded hat cocked jauntily to one side. Dolores could hardly remember a time she saw him in work clothes. "He stayed immaculate, so sharp," she'd say. If she or her sisters noticed a piece of lint on his suit jacket, they knew to pick it off him.

Dolores met Hubert when she was fourteen, soon after graduating from Betsy Ross Elementary. Her family usually went to a Baptist church across the street from their apartment, but occasionally they traveled four blocks west to the Uplifting of Humanity, a sanctified church led by her aunt Rhea. Hubert was left-handed and sang in the choir, the *dumm, dumm, dumm* of his bass drawing Dolores's gaze to him. When Hubert built up the nerve to phone and ask Dolores on a date, her parents said no. Dolores had almond-shaped eyes and a waggish intelligence, so he tried again. Her mother consented to an outing, but she calculated exactly how long it would take the two of them to travel by train to a movie theater downtown, to see the coming attractions, a cartoon, and a double feature, and then return home. "No monkeying around," she ordered. "Go see that movie, get on the El, and come back." If they returned from a date just a couple of minutes late, Hubert wouldn't walk Dolores upstairs, choosing to be unmannerly rather than face her parents. Once when Hubert bought Dolores a sweater, her father made her give it back. He had a ban on gifts—a boy would expect sex in return. "You better not be having any sex," her mother added. "But if you are, make sure to use a rubber." Her parents made it clear that there was no greater moral failing than getting pregnant, forbidding Dolores from even socializing with girls believed to be fast. When Dolores's sister became pregnant, their father forced her into an unhappy marriage.

Fearing her getting too serious, Dolores's parents didn't allow her to go steady with any one guy. Throughout high school she dated not only Hubert but also George, Clifford, Otis, Frank, and Bo. It was a short stroll from her apartment past South Parkway, which would later be renamed Martin Luther King Drive, and over to the expanse of Washington Park. She'd sit on the benches with her different boyfriends or walk with them around the lagoon, or watch the games in the fields. Dolores loved her some Frank Jenkins, but she figured she loved Hubert more. Hubert looked older than the rest of them, even though he was the same age. And he could make Dolores laugh, the way he told outlandish stories, his wit matching hers. People would say that the two of them together, with their banter, could have a comedy act on the radio. Hubert quit high school his junior year to help with his family's expenses. He was the type who'd do any sort of job, as long as it was legal. He shoveled coal into people's basements, cut ice from the lake, delivered refrigerators, laid tile.

Dolores didn't realize how well off her family was until she and Hubert started swapping tales about the Depression. More than 40 percent of all workingmen in Chicago were unemployed during those years, and the city had a shortage of 150,000 affordable homes, with the demand increasing and nothing new being built. A Hooverville formed downtown, on the outskirts of Grant Park, hundreds of jerry-built structures made of cardboard, scrap, and tar paper. "Building construction may be at a standstill elsewhere, but down here every-thing is booming," an out-of-work miner and railroad brakeman who'd been elected "mayor" of the shantytown told the press. On the South Side, Dolores had her clothes dry-cleaned at her father's shop, and when her father was drafted into the navy her mother found work at a factory making aircraft. Hubert, on the other hand, had been eating neck bones cooked every kind of way—fried, boiled, broiled, barbecued. His family had an apartment, a meager one, but they sub-sisted on the government charity boxes with NOT TO BE SOLD stamped on them.

It was no surprise to Dolores that her father didn't think Hubert good enough for her. Always the dandy, her dad would stand in the wide window of their apartment, one shined shoe propped up on the sill, and at the sight of Hubert say, "Here comes Pete the Tramp," referring to the old comic strip. He'd turn it into a kind of mocking song, as Hubert ambled up the block in his work clothes, a shovel hefted over his shoulder, a cigar stub tucked into the corner of his mouth, a dusting of coal on his hands and face.

When Dolores was eighteen, in 1947, and enrolled at Woodrow Wilson Junior College, Hubert picked her up after classes. One day when they reached Prairie Avenue he wouldn't get out of the car. He stared silently into his lap, his chest heaving. He was sweating so much he looked to be melting. Then with a dour expression he asked Dolores if she would marry him. Now she couldn't breathe, feeling she might be having a heart attack. But soon they were both laughing, and quickly picked a date for the wedding and started hurtling ahead through the years, imagining the many milestones to come in their lives together. Then a thought paralyzed Dolores: Which one of them was going to tell her parents?

"You are," he said.

"Uh-uh. You."

DOLORES AND HUBERT Wilson, with their five children in a basement apartment, never considered leaving the South Side, let alone moving to the Near North Side. They thought they had no options in the city other than the Black Belt or maybe the West Side neighborhoods that had started filling with migrants from the South around the time they got married. And besides, the wilderness of slum around the Cabrini rowhouses had its own notoriety in Chicago lore. As early as the middle of the nineteenth century, the settlement along the North Branch of the Chicago River was as undesirable a place to call home as there was in the city. It teemed with dangerous and dirty jobs. Men hauled loads between barges and trains, worked in warehouses, tanneries, meatpacking plants, machine shops, and any number of small factories. The city's first ironworks was started along the river there, in 1857; by 1870, the North Chicago Rolling Mill employed 1,500 men, producing steel rails for the tracks that were steadily traversing the country. There was a massive gasworks on the riverbank a block from where the rowhouses would be built, an endless supply of coal fed into its hungry furnaces. While the resulting gas was stored in vats, the leftover tar and coke and other effluents flowed back into the mucky river. Black clouds of soot enveloped the neighborhood at all times, which is how the district came to be known as "Smoky Hollow." The smell of sulfur was everywhere, too, and bright flames from the processed gas burst into the sky. And that's why it was also called "Little Hell."

It earned other nicknames as well, from the newly arrived

immigrants who couldn't afford to live elsewhere. When the Irish landed there, in the 1850s, the area became the "Kilgubbin," named after the section of County Cork from which the refugees of the potato famine emerged. In 1865, it was one of the city's largest "squatter villages," according to the *Chicago Times*, which wrote that the land "numbered several years ago many thousand inhabitants, of all ages and habits, besides large droves of geese, goslings, pigs, and rats. It was a safe retreat for criminals, policemen not venturing to invade its precincts, or even cross the border, without having a strong reserve force." The local Irish referred to their new home sometimes simply as "the patch." Germans followed, working as small-scale farmers, peddling their produce out of wagons. Next came Swedes, who were even poorer than their predecessors, and their numbers were so great that they supported eight different Swedish-language newspapers and the nearby stretch of Chicago Avenue was dubbed the "Swede Broadway." Then in the first years of the next century, with alarming speed, the Near North Side turned into "Little Sicily." Some 13,000 immigrants from towns in southern Italy rushed in to take over homes and storefronts from the besieged Irish, and the neighborhood quickly became the city's second-largest Italian enclave.

What distinguished this impoverished neighborhood above all else was its extraordinary proximity to Chicago's most expensive real estate. On the South Side, the Black Belt became a world unto itself. But a mere ten blocks east of Little Hell, beside the lakefront, sat the city's fanciest hotels and clubs, the high-end shops of Michigan Avenue, and the stately apartments bordered by tranquil streets of aristocratic single-family homes. One had to walk only a few minutes from this first world opulence to enter the third world meagerness along the river, with its garbage-laden alleyways, muddy lanes, and clotheslines crisscrossing overhead. The titans of industry and the city's philanthropists lived in the Gold Coast neighborhood along Lake Shore Drive. Little Sicily, less than a mile away, was where "dark, shifty eyed men with inscrutable faces lounge warily in the shadows of an area way or in the murk of a corridor," as the local press reported in 1915. The sociologist Harvey Zorbaugh made the short physical distance between these urban extremes the subject of his 1929 book, *The Gold Coast and the Slum*, in which he portrays the river district's otherness as a pummeling of the senses:

> Dirty and narrow streets, alleys piled with refuse and alive with dogs and rats, goats hitched to carts, bleak tenements, the smoke of industry hanging in a haze, the market along the curb, foreign names on shops, and foreign faces on the streets, the dissonant cry of the huckster and peddler, the clanging and rattling of railroads and the elevated, the pealing of the bells of the great Catholic churches, the music of marching bands and the crackling of fireworks on feast days, the occasional dull boom of a bomb or the bark of a revolver, the shouts of children at play in the street, a strange staccato speech, the taste of soot, and the smell of gas from the huge "gas house" by the river, whose belching flames make the skies lurid at night and long ago earned for the district the name Little Hell—on every hand one is met by sights and sounds and smells that are peculiar to this area, that are "foreign" and of the slum.

The combination of poverty and proximity helped turn the Near North Side ghetto into one of Chicago's major vice districts. The sensationalized coverage of the crimes committed there both exaggerated and contributed to the conditions. Zorbaugh describes the Italian neighborhood as a "bizarre world of gang wars, of exploding stills, of radical plots, of 'lost' girls, of suicides, of bombings, of murder." The intersection of what today is Oak Street and Cambridge Avenue, or possibly a block east at Oak and Cleveland, was shared by a cleaner's, a dyer's, a Jewish dry goods shop, and a Sicilian saloon that featured live music. It was also said to be the site of a dozen homicides a year for almost two decades. More than a hundred unsolved murders were alleged to have occurred at what came to be known widely as Death Corner. The violence in Little Sicily was attributed to the Italian Mafia, especially during Prohibition, but there was also a shadowy group known as the Black Hand. In 1908, a rumor spread that the Black Hand society had placed a nitroglycerin bomb in the basement of Jenner Elementary, the school abutting Death Corner, which was set to detonate at 2:00 p.m. There was no bomb, but in a scene that would be replayed over the next century, students ignored their teachers and tumbled down the stairwells, trampling one another, as mothers raced over from their cold-water flats to rescue their children. In their elusiveness, the Black Hand assassins were depicted as phantasms, supernatural predators. Many of the neighborhood's streets were elevated several feet above grade, and killers were said to lurk in the belowground coal sheds and half basements. In one brazen Death Corner murder recounted in the press, the culprits reportedly shot to death a man, waited for the police to arrive, and then before slipping away unseen shot the witnesses as they were giving their reports to the cops. Rumor had it that members of the Black Hand leaned over their victims and kissed them on the lips, to ensure that the ghosts of those they murdered wouldn't haunt them.

In his first year as president, in 1933, Franklin Roosevelt created the federal Housing Division, as part of the Public Works Administration. By then the shortcomings of the for-profit real estate market were evident in mass evictions and eviction riots, in homeless encampments and in countless neighborhoods like Little Sicily. "I see one-third of a nation ill-housed, ill-clad, ill-nourished," Roosevelt announced in 1937, in his second inaugural address. "The test of our progress is not whether we add more to the abundance of those who have much; it is whether we provide enough for those who have too little." In the face of what looked like a humanitarian crisis—and with decades of shaming and lobbying by progressive slum reformers and proponents of modernized housing—the government mobilized its resources. The PWA built fifty-one public housing developments over the next four years, including three in Chicago. In 1937, after two years of wrangling over the particulars, Congress passed more-extensive legislation that established a federal housing agency. Chicago and other cities formed their own housing authorities to operate the subsidy locally. The CHA, under Elizabeth Wood's leadership, picked sites for new public housing developments, and the infamy of Little Hell made it an obvious choice.

When the Frances Cabrini rowhouses were completed, in 1942, hundreds of federal, state, and city officials attended the dedication ceremonies on Chestnut Street and Cambridge Avenue. Mayor Edward Kelly, "his red hair like slag in a sea of Mediterranean complexions," as one reporter recounted, announced that the 586 units of public housing "symbolize the Chicago that is to be. We cannot continue as a nation, half slum and half palace. This project sets an example for the wide reconstruction of substandard areas which will come after the war."

Father Luigi Giambastiani, of Saint Philip Benizi Parish, located next to Death Corner, was among the leaders of the neighborhood's Sicilian community to suggest that the rowhouses be named for Mother Francesca Cabrini. A nun who settled in the United States in 1889 and worked among the Italian poor, Mother Cabrini established a school and a hospital in Chicago as well as more than sixty institutions countrywide. She died in Chicago in 1917. "To you she is a social worker," Father Giambastiani told the press, "but to us she is a saint." In 1946, Rome would agree: Mother Cabrini became the first American citizen to be canonized, the patron saint of immigrants.

Working families with young dependents were initially given preference for admission into the planned Cabrini rowhouses, since they were often rebuffed in the open market. To secure a coveted spot, married couples had to pass rigorous screening and earn enough annually to meet minimum rent requirements. Very poor families, those who were unemployed, unstable, or unseemly—the new public housing wasn't intended for them. The subsidy wasn't charity or humanitarian assistance; the developments were supposed to revitalize the slums, not replicate them. Days before construction on the Cabrini rowhouses was to get under way, however, the Japanese bombed Pearl Harbor. Many of the factories in and around Little Sicily were converted to war-related industries. Amid the city's affordable housing shortage, the CHA agreed to give families of veterans and war workers priority in the new development; in exchange, the agency was able to secure rationed building supplies and proceed with construction. The renters would pay between $24 and $37 a month for a three-bedroom home, based on their income, with electricity, hot water, and heating fuel included. But to accommodate war workers with their higher salaries, the CHA more than doubled the maximum amount that residents could earn yearly and still qualify for a unit, from $900 to $2,100. This at a time when a third of all Chicago families, and just about everyone in Little Hell, had an annual salary of less than $1,000. They would likely not find a home in this "Chicago that is to be."

DOLORES WILSON

BEFORE DOLORES AND Hubert were married, Hubert was picked up by the police. He was accused of robbing a cleaner's on his South Side block. His family assured the police that he'd been at home with them, but a neighbor fingered him as the thief, saying she thought the guy she may have seen in the dark alley three stories below looked sort of like Hubert. That was enough for the cops. They moved him from one precinct house to another, so his relatives couldn't find him. At one of the police stations, officers handcuffed his wrists to the arms of a chair, demanding that he confess to the crime he didn't commit. When he refused, two white cops standing on either side of him counted down, and at the same moment each one bashed an ear.

The headaches from that beating lasted for the next thirty years of Hubert's life.

Sometimes Dolores would think about all the pain that the police caused folks, how Chicago cops abused their power, especially in the city's black neighborhoods, and she'd shake with contempt, losing her ability to speak. There was a black officer in their neighborhood named Sylvester Washington, though everyone called him "Two-Gun Pete," for the pair of pearl-handled .357 Magnum revolvers he wore around his belt like an Old West gunslinger. Once, when Dolores had taken a group of children to the Bud Billiken Parade, an annual event in the South Side sponsored by the *Defender*, she lined them up to buy ice cream. Two-Gun Pete pushed over the little boy at the front of the line, for no other reason than to see the children topple in a row like dominoes. One night, Hubert was on his way home, wearing sunglasses, when Two-Gun stopped him. "Is the moon too bright for you?" the cop demanded. He smacked the glasses off of Hubert's face and ordered him to pick them up. When Hubert bent over, Two-Gun kicked him in the back. Then the cop slapped the shades off of Hubert's face again and told him to get them. The police officer had, officially, killed nine people while on duty, and the department, rather than punishing Two-Gun, rewarded him with promotions. Hubert knew what was coming, but he didn't want to add to Two-Gun's body count. The kick sent him to the ground.

With Hubert locked up for the cleaner's robbery, his family hired a lawyer, and that put an end to the railroading between holding cells. Dolores baked biscuits and sent them to the jail. As Hubert bit into each one, he'd find tiny pieces of paper stuck to his tongue, notes that read, "I love you." It turned out that it was Bo, one of Dolores's other suitors, who'd broken into the cleaner's. He was arrested, and Hubert was cleared of the crime.

The couple named their first child after Hubert, but people called little Hubert "Chuck-a-Luck" as a baby, and when he took his first steps Dolores's brother cried, "Che Che," and that's the name that stuck. Then came Michael, Debbie, Cheryl, and Kenny. Two of them were born in their apartment, with a county doctor showing up only after the delivery to cut the cord. Dolores and Hubert wanted better for their growing family than their basement apartment, so they tried to purchase a home. They learned about a subdivision soon to be built on the distant South Side. It was being constructed from the ground up, on empty land. With the help of Dolores's aunt, they made the down payment for the house that existed only as an architectural plan. Dolores's sister and sister-in-law and their families also bought into the same development. They each picked out their plots, and in their anticipation they talked endlessly together about what the houses would look like—how you'd walk through the front door and step down to enter the living room, like in a television program. Then Hubert came down with the Asian flu. He was working in construction, and for two days he couldn't lift himself out of bed. On the third day, before the fever broke, he forced himself to return to the job, but it was too late. His boss fired him. The Wilsons fell behind on their payments and eventually lost the property. Dolores's sister and sister-in-law soon pulled out, too. Losing the down payment didn't bother Dolores too much, since she never got overly excited about money. But losing the dream of the thing hurt. She'd eventually shrug and say about the flu that ended their chance at owning a home, "How come some other part of the world always got to mess you up?"

The question for Dolores wasn't whether she wanted to move into public housing but which of the new developments would take her. "Public housing was my best bet for my children," she'd say. There weren't any vacancies in the Ida B. Wells Homes, a South Side complex that had been built exclusively for black residents. When the Wells Homes were completed, in 1941, 18,000 families applied for the 1,662 units; First Lady Eleanor Roosevelt visited, and the opening was cause for a parade, with marchers waving placards declaring "Better Housing, Better Health, Better Citizens." Dolores filled out a CHA application at a church on Forty-Seventh Street, in the Bronzeville neighborhood, and a social worker asked for documentation showing that she was married and that Hubert had a job. "You had to prove everything," Dolores would recall. The interview covered their rental and work history, Che Che's grades and the family's values. Another person from the agency visited their apartment to see how Dolores kept house. She had no idea what they wrote down in their notebooks. She cared only that she passed the test. The elaborate screening worked as a kind of validation; they, among the city's hordes of working poor, had proven worthy of elevation. In the new developments, the CHA awarded prizes for the best gardens and issued fines for littering. A spot in public housing felt like a leap into the middle class.

She first thought she might try for Altgeld Gardens, on the far South Side. It was built during the war, a hundred and thirty blocks from the Loop, so that African Americans could work in the nearby steel mills along Lake Calumet. With its 1,500 homes on 157 acres of formerly vacant land, Altgeld was like a self-contained, all-black village. The public housing development was so large and isolated that it had its own on-site Board of Health clinic, library, and church, its own grocery and drugstore, its own nursery, elementary, and high schools. One of Dolores's aunts lived there, in a compact box of a rowhome with a little upstairs and downstairs, a tiny front lawn, and a backyard. Her aunt loved it, and Dolores was drawn to the idea of tending her own garden. In her basement apartment on Prairie Avenue, they hardly got any light at all, and she didn't dare open the window too wide for fear of what might blow in from the alley. But Dolores couldn't get over how remote Altgeld felt. It was basically in Indiana. It didn't connect to streetcars or trains, and the bus service was spotty. "When I'm ready to go somewhere, I don't want to have to wait for a scheduled bus that might not be there for an hour," she'd say.

That's when she looked at Cabrini. After its 1950 survey of the Cabrini Extension Area, the CHA razed the slum alongside the rowhouses, demolishing the 2,325 units of substandard housing there; in their place, the agency erected fifteen separate towers that stood seven, ten, and nineteen stories tall, a total of 1,925 apartments. The original plans called for sixteen buildings, with the tallest of them reaching only nine and sixteen stories, but to further cut down on costs, the extra floors were added and an entire building eliminated. The nearly identical designs of the high-rises added to the speed and inexpensiveness of the construction. The towers were stripped-down, modernist plinths, with row after row of windows. Constructed of wine-colored bricks set inside a latticed frame of exposed white concrete, they had the look of graph paper, with repeating red boxes outlined in white. Because of the brick color, the fifteen high-rises came to be known locally as the "Reds."

Although the high-rises resulted in a net loss to the area of four hundred units of housing, the complex was prized for using just 13 percent of the thirty-five-acre site. The towers were ringed by vast plazas and lawns. Builders trucked in 25,000 cubic yards of topsoil; they planted 10,000 bushes and 500 trees, and protected it all with 23,000 square feet of chain-link fencing. The several square blocks between Chicago Avenue and Division Street were closed off to through streets and re-formed into a massive, pedestrian-only "superblock," with no businesses, street traffic, or other uses apart from housing, a school, and a community center. It was a purity of modernist city planning, influenced by the avant-garde "towers in the park" urban reimagining of the Swiss-French architect Le Corbusier. When all fifteen of the towers were fully operational in 1958, the Cabrini Extension was Chicago's largest public housing high-rise complex, a template for the many others to come. "We thought we were playing God in those days," said Lawrence Amstadter, one of the development's architects. "We were moving people out of some of the worst housing imaginable and we were putting them into something truly decent. We thought we were doing a great thing, doing a lot of innovative design things."

A handbill for the new development was written as if speaking directly to Dolores. "For families for whom private enterprise has not been able to build or operate decent, safe and sanitary dwellings at prices within low-income budgets . . . just a mile, or 10 minutes by street car, from the Loop." It was close to "Near North Side parks and beaches . . . handy to swimming and fishing in Lake Michigan," and "down the street from Montgomery Ward's big retail store." The Larrabee Street bus offered a straight shot into downtown. You could walk to a hundred different factories to find work. The Cabrini Extension towers weren't designed with interior hallways; rather, each floor included an open-air gallery. "Sidewalks in the air," Elizabeth Wood called these exposed ramps. In a high-rise with hundreds of apartments, each family could have "ready access to the out-of-doors," the CHA pointed out. "An open porch on the nineteenth floor is a convenient play space for small children under mother's watchful eyes. . . . It adds zest to living in the new home for families who formerly had to come out of basements to see daylight."

And the CHA promised that the addition of nearly two thousand units of public housing to the six hundred that already existed in the rowhouses would be a boon to the long-neglected area. More public housing meant more modern construction, more retail, and more amenities and services. The CHA advertised that the tall apartment towers would "blend with the rowhouse and garden apartments of present Cabrini," declaring, "Twenty-five hundred families in housing easy to make into good homes will change the whole neighborhood!" Everyone in Dolores's family lived on the South Side. But the idea of life at Cabrini grabbed ahold of her. "Near North, near everything," she recited like a hymn.

2

The Reds and the Whites

DOLORES WILSON

THE WILSONS WERE one of the first families, in 1956, to move into their nineteen-story Cabrini Extension high-rise. They were assigned an apartment on the fourteenth floor, and everything smelled of fresh paint. Dolores had never been that high in the sky before. She forced herself to the edge of their "sidewalk in the air," the walkway outside the apartment, and clutched the chest-high fencing as she peered down. The cars looked no bigger than toys. "I almost fainted," she would recall. That was the first day. "Then after a while you grow used to everything." Soon she was delighting in her lordly view of Chicago's rippled skyline and the blue-gray blur of the lake disappearing into the horizon. Just as the CHA had advertised, she stretched out on what they called the ramp, enjoying the breeze, while her children played alongside her. Flies and mosquitoes didn't reach fourteen stories, nor did all the sounds of the street. She started to feel sorry for her neighbors who filled the apartments on the floors below.

Like all the new Cabrini towers, the Wilsons' building was known impersonally by its address, 1117 N. Cleveland, which was painted in a somber blocky script above the front entrance. The architect Lawrence Amstadter had wanted to install metal numerals and letters, explaining that it would actually cost less to do so. But he was told the metal gave off the appearance of being pricier. "Public housing was considered charity," he said. "It had to look economical." The subsidy had the tricky task of ensuring that those with too little received only just enough. That was fine by Dolores. "It was the projects," she'd say, by which she meant it was clean and safe and spectacular to behold. They now had five full rooms, with a large living room, a kitchen, and their own private bathroom. She and Hubert slept in one bedroom, the boys in another, and the girls in a third. There was a fridge, a stove, and hot and cold water. The walls were plum, the floors and ceilings unbreached.

Her mother would phone when the temperature dropped into the teens, worried that Dolores might be inside the apartment barefoot. Dolores would hush Hubert as he shouted from the other room that she didn't have on any shoes. But this wasn't a South Side tenement. The heat in their building pulsed through the floors and enveloped them. Outside, the frigid air rolling in off the lake or the prairies blasted the high-rise. But inside the temperature exceeded eighty degrees. On the most penetrating winter days, with ice layering on the ramps and the wind roaring like a jet engine, Dolores still cracked her window. And they paid just a small, fixed amount for utilities and a monthly rent that was based on their annual income. The interior walls, too, were solid cinder block. If a fire broke out in any kitchen in the building, the flames couldn't go any farther than that apartment. Dolores would demonstrate by entering each room and shutting the door. "It was fireproof," she would say. "The whole thing was fireproof. Even the smoke couldn't get in there. It was like heaven."

The comparison turned out to be a common one: something as elemental as a modest home seemed divine after the damnation of its absence. "It's heaven here," a different mother who moved into one of the other newly built red Cabrini high-rises told the press. "We used to live in a three-room basement with four kids. It was dark, damp and cold." J. S. Fuerst, one of Elizabeth Wood's "do-gooders" at the CHA, headed up the agency's research and statistics division and had overseen the 1950 survey of the Cabrini Extension Area. For a book

of interviews he titled *When Public Housing Was Paradise*, Fuerst
collected dozens of similar testimonials from early occupants of Chicago's public housing. "It's almost like I died and went to heaven," one
tenant said about moving into a low-rise development. "We felt it was
just paradise," another resident told him. "We felt this was just the
greatest housing that we could live in!" In Dolores's basement apartment on Prairie, if the shared toilet didn't flush or a circuit shorted,
she could call her landlord. But he might just as soon put them out
as fix the place. At Cabrini, the government owned her home. There
was a city agency responsible for answering her requests. Her building had a team of janitors on call around the clock. Groundskeepers
maintained the gardens and lawns that circled her tower like a moat.

Dolores discovered, too, that she liked living around a lot of
people. Her building and its conjoined nineteen-story twin, 1119 N.
Cleveland, together held 262 apartments and almost a thousand residents. Towers of the same crosshatched red brick faced her on all
sides. "The more the merrier," Dolores would say. "It was nineteen
floors of friendly, caring neighbors. Everyone watched out for each
other. No gangs, drugs, or shootings." One of her sisters moved to a
residential street on the West Side, and it was so quiet a cat slinking
in the bushes would scare Dolores. "If everybody is laughing and
happy or fighting, I know there's life out there," she'd say.

All the families in the high-rises had gone through the same
careful screening, and most of the households had two parents at
home. People kept their doors unlocked and dropped by one another's
apartments when they needed to borrow sugar or a cup of milk. They
looked after one another's children. Dolores's next-door neighbors
were Puerto Rican, and she got along with them great, even though
the aromas of their cooking wafted onto the ramp. "I like garlic," she'd
say, "but I don't like the smell of it." Dolores became close friends
with a woman named Martha who lived on the second floor and had
five children as well. They'd go on outings together, sometimes bringing along another girlfriend from the building who also had five kids.

They'd play in the large playground. Or they'd walk to Seward Park or the Isham YMCA or to Pioneer market, the three adults marching the fifteen children. "We looked like a parade," Dolores said.

In the mornings, thousands of people flushed out of the high-rises on their way to work. Many had jobs at Montgomery Ward, the giant retailer and mail-order cataloguer, which was the neighborhood's largest employer. Oscar Mayer, an immigrant from Bavaria, had opened his business selling "old world" sausages and Westphalia hams in Chicago in 1883, and five years later moved to the corner of Sedgwick and Division, just across Seward Park from Dolores's building. The eight-story plant produced hot dogs and sliced lunchmeats. Dozens of other small factories near the river made everything from Turtle Wax and tamales to tractors, donuts, comic books, and children's toys. Industries lined the neighboring boulevards, producing paint, radios, elevator parts, billboards, and Dr. Scholl's shoes and arch supports; others manufactured clothes, luggage, cameras, power plant equipment, picture frames, auto parts, and office supplies. A number of the Midwest's candy companies made their confections in the area, and Dolores and her children could taste the chocolate in the air. The neighborhood had an old settlement house, funded by the city's welfare council, called the Lower North Center, which had been housed in an eighty-year-old building. When the Cabrini Extension went up, it was rebuilt into a low-slung brick building the same color as the Reds, with classrooms, a nursery, meeting halls, and a gymnasium. CHA tenants used sewing machines there to learn dressmaking, and they went to the center to get coached for the civil service exam.

Hubert held a number of different jobs, and quit a number of them, too. "Why are you home so early?" Dolores would ask him, knowing the answer. "I just walked out," he'd reply, explaining that they'd treated him unfairly. But the next morning he'd head out and land something new. One of their neighbors had walked into the Seeburg jukebox plant the same day he'd moved into his Cabrini apartment and was offered a position. "When do you want me to start?" he asked. "Right now," came the response. Only once, after Hubert had been laid off from a seasonal construction job, did the Wilsons consider going on welfare. But they thought that they'd be forced to give up their car and television and other belongings before they qualified. Instead they borrowed a little money from Dolores's parents, just enough to get by, and Hubert soon found work again. Eventually he was hired by the CHA as a janitor. He joined the brigade of custodians moving each day among the fifteen Cabrini high-rises, hauling trash, cleaning hallways and stairwells, and straightening up anything that was in disarray.

ONE OF DOLORES'S new neighbors was also among Cabrini's most famous residents. Jerry Butler, the soul singer known as the Iceman, had been born in Sunflower County, Mississippi, in 1939, and at three his parents left sharecropping in the delta for work in the war industries in Chicago. They settled into a tenement just a block from the Frances Cabrini rowhouses, the family living in a three-room basement apartment with no hot water that sat squarely within the area detailed in the CHA's *Portrait of a Chicago Slum*. The law now gave priority in the new high-rise public housing to families cleared from slum sites, and after the Butlers' building was torn down to make way for the Cabrini Extension towers, they moved into 1117 N. Cleveland. While a student at Jenner, the local public school, Butler delivered newspapers on the Gold Coast, and at age twelve he had a job in a plastics company, operating an injection-molding machine from 4:00 p.m. until midnight. As a teenager, he also stuffed mattresses at a nearby factory, and one of his coworkers took him to the West Side to sing at a basement ministry called the Traveling Souls Spiritualist Church. The minister was a mystic who said she communed with a spirit guide from among the dead. She also had a grandson, a pint-size nine-year-old named Curtis Mayfield, who knew how to play the guitar and piano and who could sing in a high tenor that complemented the lower register of Butler's already silky baritone. The kids formed part of a quartet, the Northern Jubilee Gospel Singers, and began to tour spiritualist conventions. A couple of years later, Mayfield moved with his mother and siblings into a Cabrini rowhouse on Hudson Avenue. With its front- and backyard and private toilet, it was luxurious after the run-down hotel where they'd been staying.

Jerry and Curtis knew of Ramsey Lewis, who had grown up in the Cabrini rowhouses and had already gone on to cut two records with his jazz trio. But they weren't into jazz. They sang doo-wop together outside the towers. They put on shows at the Lower North Center and practiced in the basement of Butler's high-rise and in a club room at the Seward Park field house. By the benches in the park, a wino named Doug played an old guitar, producing a masterful sound, and the boys studied him for hours. Butler wanted to be a chef then, and he crossed Division Street, two blocks from his apartment, and signed up for culinary classes at Washburne Trade School. Many of the city's unions operated apprenticeship programs out of the high school, and professional chefs sometimes showed up to hire for their kitchens. When Butler started as a freshman, he was one of only a handful of black students. Then as the school became fully integrated, the unions pulled their apprenticeships.

Jerry sometimes bussed tables as well at his uncle Johnny and aunt Pearlie's restaurant, next to Seward Park. Packed from morning through night with everyone from Irish cops and numbers runners to bus drivers, judges, and ministers, it was like a community round-table for the ever-changing neighborhood. People his uncle knew from back home in Mississippi would make the journey up to Chicago, and he'd give them a job in the restaurant until they settled into something steadier.

By then Italians and African Americans had been living side by side in the area for decades, mostly peaceably but sometimes not. In 1935, an Italian property owners' association took it upon itself to try to evict 4,700 black renters from the Near North Side. Father

Giambastiani, of Saint Benizi Parish, defended the illegal acts: "The landlords were protecting their property values as they had a right to do." Although the Sicilians he spoke for were manual laborers or shopkeepers, they'd come to the city and embraced the American dream of owning property. The rate of home ownership for first-generation Italians in Chicago was twice that of native-born whites. When the CHA was clearing the Little Hell slum in the forties to make way for the Cabrini rowhouses, a group of Italians refused to sell. They didn't want to sacrifice personally for the government's notion of the greater civic good. Initially, the Cabrini rowhouses were to cover fifty-five acres, extending all the way from Chicago Avenue to Division Street. But the CHA had to scale the development back to sixteen acres when it couldn't secure the land.

The chairman of the CHA under Elizabeth Wood was an African American architect and businessman named Robert Taylor. One of the few black officials in city government, he had helped found a savings and loan on the South Side, the rare banking establishment in Chicago to offer home loans to minority borrowers. He shared Wood's belief that the integration of public housing was a practical necessity as well as a moral one. A quarter million Chicago families lacked safe and sanitary housing, with the greatest demand coming from the city's segregated black neighborhoods. Federal law dictated that new public housing developments couldn't change the existing racial makeup of a neighborhood, only match it. The CHA had at first avoided the trickiness of this "neighborhood composition rule." Two New Deal–era sites inherited by the agency were on vacant land in all-white neighborhoods, and a third included a smattering of black families relegated to a corner of the complex with its own separate stairwell.

When the Cabrini rowhouses were completed, the surrounding area was 80 percent white and 20 percent black, and so the small development held to that quota. Black and Italian children played together in the parks and at the YMCA. They attended the same schools. Neighbors of different races joined up for block parties and the huge feast-day celebrations of Saint Dominic and Saint Benizi. "With an integrated project we were all one big family," a white rowhouse occupant recollected in *When Public Housing Was Paradise*. "It was a real village." The *Defender* covered the opening of the rowhouses with uplifting images of middle-class domesticity. A black family of four enjoying a first meal in their dining room. Two Cabrini mothers, one white, one black, in similar blouses and long skirts, their shoulders touching as they gaze into an oven; the caption explains that they are joined for a child's twelfth birthday and that the black mother "is probably giving advice on how the birthday cake should be cooked, or perhaps she is simply kibitzing as the women will do."

Father Giambastiani delivered a prayer at the Cabrini rowhouses' dedication. Within weeks, however, he was writing letters to Elizabeth Wood, complaining that the social experiment there had failed because blacks were being placed "on the same level and house to house with the White people: this is being resented by all and I must add, in order to be candid with you, that I don't like it either." Italian property owners lobbied the housing agency to segregate African American renters to the southwest corner of the development—"to preserve the character and values of the neighborhoods on the north side." Giambastiani began to advertise the parochial school of Saint Benizi as an all-white alternative to the integrated Jenner; the neighborhood's two public parks set up separate hours for white and black children to use the facilities. Some in the community revived the terror of their parent's childhoods, calling themselves the Black Hand and attacking their neighbors. In April 1943, a group of whites shot into a Cabrini rowhouse occupied by a black family, and several hundred people took to the streets as fights broke out. During the next months, more police were stationed at Cabrini than at any other Chicago housing development, as city officials feared that the Near North Side might erupt into a full-fledged race riot. The *Defender*, hopeful, reported that amid the turmoil two hundred children at the rowhouses elected a fourteen-year-old black youth to head their junior government, the headline declaring, "White Kids Rebuff Hate, Elect Negro Boy 'Mayor.'"

The neat columns of the Cabrini rowhouses were bordered to the south by Chicago Avenue and to the west by Montgomery Ward's gargantuan, two-million-square-foot warehouse that curved where the Chicago River bowed. Directly to the east, several of the seven- and ten-story Cabrini high-rises were positioned like chess pieces on a single tract of land hardly bigger than a square block; across Oak Street to the north, nineteen- and ten-story towers rose up like a wall. The Reds lined Larrabee heading north, and Division going west, and Cleveland heading back south to Oak, completing a four-sided bulwark that formed around the parkland between the buildings. Many rowhouse residents saw the high-rises not as an extension of their community but as something separate. The distinctions weren't merely design related.

By 1950 the population of the twenty-five blocks cleared to make way for the Cabrini Extension towers had jumped to 80 percent black, as older Italians moved out and younger families moved in. The CHA touted one of the new Cabrini high-rises as its "International Building," reporting that some 262 African American, Armenian, Chinese, Danish, Eskimo, German, Indian, Irish, Italian, Mexican, Polish, Puerto Rican, Scottish, Swedish, and Turkish families live there "in harmony, in friendship and in peace." But overall, about 90 percent of the 1,900 families who moved into the fifteen high-rises were African American, the majority of them like the Wilsons and the Butlers—two-parent, working-class, and desperately in need of adequate housing. In Dolores's building, she knew of only one white family; the wife's name was also Dolores, and she used to sell her blood at Mount Sinai Hospital to make a little extra money.

Unsettled by the shifting demographics, many white renters quit their rowhouse apartments, with a great number of them moving to the western suburbs. "I saw the atmosphere go down in Cabrini," said a white mother who lived in the rowhomes with her family until the first high-rises opened. "I saw it change from sort of an ideal little community into a place that you wanted to get out of." The CHA no longer gave war workers or veterans preference in the developments, and the agency felt compelled to push out families who exceeded the reduced income limits. "Be proud to move out, so that a lower-income family can have the advantage that you have had," Elizabeth Wood told better-off families during their evictions. But the agency had difficulty finding other white tenants to replace those leaving the Cabrini rowhouses. The CHA had upped the quota of black residents allowed at the low-rise development, though only slightly, and it submitted to the demands of other occupants and kept sections of the rowhouses solidly white. Wood even had the agency print up tens of thousands of applications for the Cabrini rowhouses and distribute them among potential white residents. Brochures featured white children playing in yards and on the swings outside low-rise apartments, a picture of suburban comforts at inner-city public housing costs. But the efforts weren't enough. Whites made up less than half the rowhouse population by the early fifties. Units sat empty, while hundreds of black families who had passed the CHA's screening weren't considered for them. Amid a citywide housing shortage, Wood and Taylor found themselves defending a quota system that seemed unethical if also necessary. Wood believed, not inaccurately, that there was a racial tipping point at which whites would view most inner-city public housing as a subsidy meant solely "for Negroes." By the start of the 1960s, white families occupied only forty-two of the nearly six hundred units in the Cabrini rowhouses.

The Butlers were eventually forced to leave their apartment in 1117 N. Cleveland. Jerry and Curtis Mayfield formed a band with a couple of friends who'd made the journey to Chicago from Chattanooga, Tennessee. As the Roosters, they were performing at a talent show at Washburne Trade School when a promoter named Eddie

Thomas heard them and promised he could make them into stars. Thomas explained that the group's name sounded too country. They settled on the Impressions. *Always leave an impression.* In 1958 they recorded their first song, on the South Side's Record Row. Butler had written the tune a couple of years earlier, when he was a teenager gazing out his high-rise window, a kid pining for idealized love. It had no hook and no real sing-along parts. "A poem set to music," he'd say. "For Your Precious Love" reached number eleven on the *Billboard* charts. The young men played the Apollo Theater in New York, and Butler received a royalty check of a couple hundred dollars.

The CHA soon contacted the mother of Jerry "the Iceman" Butler, saying the family's joint income exceeded the maximum allowance for public housing. The Butlers moved to a house on the South Side. Curtis Mayfield stayed on in the Cabrini rowhouses for several more years, selling cigars in office buildings in the Loop before his own music career took off again and he could afford an apartment of his own in a new high-end complex just ten blocks away.

WHEN THE HOUSING Act of 1937 was being debated, it was opposed by every real estate trade group, by builders, suppliers, US Chambers of Commerce, property owners' associations, and the Departments of the Interior and the Treasury. Eight years earlier, the department store scion Marshall Field III opened one of Chicago's first subsidized housing developments a couple of blocks north of the future Cabrini rowhouse site. The Marshall Field Garden Apartments included 628 low-income dwellings as well as a progressive demonstration school, workrooms, and storefronts on the ground floor. But the entire complex was privately funded, the charity intended also to attract additional commercial and residential development to the surrounding Italian neighborhood.

State-run housing was seen as something altogether different. Critics of the day denounced it as government overreach, with the handouts maligned as anticapitalist, socialist, and generally un-

American. A public housing development, the very idea of it, clashed with the country's exalted sense of home ownership, with a national ethos wrapped up in visions of the frontiersman, the log cabin, and the self-made entrepreneur. Although the subsidy was reserved only for stable families with modest incomes—the "deserving poor"—the ceiling on what residents could earn was said to discourage hard work, acting as a sap on initiative and pluck. Ayn Rand's *The Fountainhead*, from 1943, dramatized the backlash against Roosevelt's call for a deeper social contract of shared responsibility. The hero of the novel is an architect of a public housing complex who becomes outraged when he returns from a trip to discover that his bare-bones high-rise has been compromised to include "the expense of incomprehensible features." Among what he deems nonessential aesthetics are blue-painted metal balconies ("You gotta give 'em a place to sit on in the fresh air," he's told), a gymnasium demanded by the social worker in charge of tenant selection, extra doorways and windows, an awning, decorative brickwork, and a carved relief sculpture above the building's main entrance. In an act that is portrayed as a valiant defense of his convictions, the architect dynamites the entire building.

Like all the New Deal programs, public housing wasn't positioned as a renunciation of the for-profit capitalist system; instead, it was a massive government intervention meant to save it. Hiring under the Public Works Administration, relief for farmers, aid to senior citizens through Social Security, food stamps for the urban poor, and public housing—these reforms treated poverty not as a personal moral failing but as a widespread social and economic injustice that the country was obligated to right. The programs were also intended to jump-start the economy, creating jobs and resuscitating the industrial sector through large-scale civic construction projects. In 1949, Congress passed legislation that promised to fund 810,000 additional units of public housing. President Harry Truman, in signing the bill into law, declared that it was the nation's collective duty to provide "a decent home and a suitable living environment for every American family." But it was voted into law only after being packaged as part of an expansive urban renewal bill that promised to revive cities and rebuild their infrastructure. In the House of Representatives, the last of several efforts to strip the bill of its public housing provisions came within five votes of passing. Catherine Bauer, the crusader who helped will the nation's public housing program into being, would regret that the subsidy was "continuously controversial, not dead but never more than half alive."

The CHA, which campaigned to get 40,000 of the new public housing units built in Chicago, argued that modernizing the worst neighborhoods would not only benefit the poor but also restore the city to profitability. "A Chicago slum district costs the taxpayers six times as much annually as is paid back to the city in taxes," the agency advertised. The CHA estimated that it could build as many as 150,000 units of public housing without taking a single customer away from the standard private-sector housing market. In a further appeal to self-interest, the agency added that quality low-rent homes would keep the poor from seeking better housing elsewhere, such as in *your* neighborhood.

Maybe most tellingly, the same Depression-era legislation that funded the country's first public housing developments also created the federally insured private home loan. With a Federal Housing Administration guarantee, banks were able to offer large home mortgages to people, paid off in small increments over an unprecedented thirty years, allowing families to put down as little as 10 percent of a house's cost. With this revolution in home financing, families could afford bigger and better homes, and developers and designers began to build homes in anticipation of the growing demand. The subsidies to the speculative real estate market dwarfed the outlays to public housing and to any programs offering rental assistance. Here was the American dream in practice: a select number of the nation's ill-housed got public housing in cities; those better off were able to buy homes, increasingly in the suburbs, with the government taking on the risk. African Americans—who'd been owned as property a little more than two generations earlier and then had their own property systematically taken away from them in the Jim Crow South—found themselves excluded from a gamed mortgage system that allowed others to accumulate wealth with little investment and with minimized risk.

The most virulent opposition to public housing in Chicago came from white working-class families who felt trapped in the racially changing city. With the flow of migrants from the South surging after the Second World War, Realtors used shady blockbusting tactics, surreptitiously moving a single black family into an area, then drumming up fears among white neighbors of decreasing property values and a growing threat of crime, to buy up their houses on the cheap. Many white Chicagoans took advantage of the new federally insured mortgages to speed out to all-white suburbs. Those who couldn't afford to move were desperate to preserve whatever tenuous hold they had on their neighborhoods. Their parents or grandparents had come to Chicago from Europe, and they'd made a home in the city, buying a bungalow among families much like themselves. They believed all of that was now in jeopardy. There were nearly five hundred outbreaks of inter-racial violence in Chicago between 1945 and 1950, with 350 of them directly tied to fights over housing.

At the CHA's Airport Homes, near Midway Airport on the West Side, white residents from the neighborhood vandalized the public housing development in 1946 at the mere rumor that black World War II veterans might be allowed to live there. Several men strongarmed their way into the housing office, stole keys, and moved white families into vacant units. When the CHA did rent one of the subsidized apartments to a black family weeks later, the husband was a decorated veteran, one of the 175,000 returning from the war who applied for government housing assistance in Chicago, and he was selected as carefully for the occasion as Jackie Robinson was for his integration of Major League Baseball the following spring. As a

precaution, the CHA brought in the family during the day, when most
men from the community were away at work. But white women took
up the charge, hurling curses and dirt and bricks. Similar protests
occurred the following summer when the CHA tried to integrate the
Fernwood Homes, except that the mobs in the South Side neighborhood of Roseland were even larger, reaching five thousand. White
youth pulled black drivers from passing cars and beat them. Local
churches and community groups endorsed the violence that continued for weeks, and the neighborhood's alderman condemned not the
perpetrators but the CHA for including Roseland in these "ideological experiments."

The largest of the city's public housing race riots occurred at the
Trumbull Homes, in the South Side neighborhood of South Deering.
In 1953, a light-skinned black woman applied for a public housing unit in Chicago, and the housing authority, assuming she was
white, accidentally integrated the development of three-story apartment buildings and rowhomes. South Deering was a working-class
area of Italians, Poles, and Slavs south of the encroaching Black Belt
but north of all-black Altgeld Gardens, separated from it by Lake
Calumet and the surrounding steel mills. When the woman showed
up with her husband and children, the resulting protests lasted for
months, with locals setting off homemade bombs and crowds burning
down businesses. A community newspaper wrote about the "savage,
lustful, immoral standards of the Southern Negro," submitting that
South Deering might consider accepting black residents only after
black people were civilized; in the meantime, the white neighborhood would defend itself, while other "spineless communities" were
"raped, robbed, and murdered."

In 1947, amid the ongoing racial upheaval, the city's Democratic
machine forced out Mayor Edward Kelly. Kelly had allowed Elizabeth
Wood to keep the CHA free of the machine's patronage hires and
preferential contracts, and he tolerated an agency agenda often at
odds with the interests of local ward bosses. The new mayor, Martin

Kennelly, was picked by the regular Democratic organization because he could be counted on by the city's power brokers. Some progressive housing reformers at that time believed it was preferable to build new public housing on vacant land on the outskirts of the city, where the net gain of homes would be greatest and the poor could start anew; others insisted on building atop existing slums, so the city's most deplorable housing would be replaced. Elizabeth Wood and Robert Taylor wanted to spread around new sites, apportioning the working poor across a number of different neighborhoods so that the entire city shouldered the responsibility of housing the disadvantaged. But under Kennelly, the city council was given the final say on the selection of public housing sites. Spurred on by the racial fears of their constituents, aldermen rejected every site proposed by the CHA that was in a white neighborhood. When the council debated the issue for four consecutive days in 1950, hundreds of residents from white areas were bussed in to fill the gallery. Of the 12,500 new units to go up in Chicago, just 2,000 of them would be constructed on empty land; the remaining 10,500 were built in black slums or as extensions of existing public housing developments. The CHA reluctantly accepted these terms. Any housing was deemed better than further delays or no new homes at all.

From that point on, the agency no longer bothered to suggest sites for its developments beyond existing ghettoes. Robert Taylor, humbled, resigned. And while Wood stayed on until 1954, when she was forced out, she had been stripped of her independence. "I was so undercut politically that I was feeble, I was floundering around," she said later. "We no longer had power to select where the projects were going to go, and we had very little space to work with, so we had to go to the high-rises."

Of the thirty-three housing projects built in Chicago in the coming two decades—a total of 168 high-rise buildings—all save one went up in an overwhelmingly black neighborhood or an area far along in its transition from white to black. Public housing still

represented a major upgrade for families in these decaying neighborhoods. Yet large public housing complexes now had the effect
of solidifying the boundaries between black and white Chicago in
superblocks detached from the street grid and in towers of concrete
and steel. In Arnold Hirsch's history of this pivotal reconfiguring of
the city, *The Making of the Second Ghetto*, he calls the site selection
of Chicago's high-rise public housing a "domestic containment policy." Wood complained at a public forum soon after her firing, "It is
the will of politicians to keep Negroes where they are."

DOLORES WILSON

DOLORES ALWAYS CLAIMED she didn't want to take on any additional responsibilities, that she preferred to be a spectator. But that
just wasn't true. In her own subdued way, she'd end up running six
different committees. She showed up for the PTA meetings at Jenner,
her children's school, and found herself elected treasurer and later
president. She served on her building council, planning activities for
the teens and younger children in the tower. The tenants celebrated
the "birthdays" of their high-rises at Cabrini, the date coinciding
with their address, and Dolores helped throw a big party in the 1117
N. Cleveland rec room every November 17. She attended Holy Family
Lutheran, a small cube of a church across the field from her apartment, joining its board, and started volunteering with an organization that helped the formerly incarcerated manage their return to free
society.

She spent many hours as well at the Lower North Center. Her
children attended camp there and used it for dance and music classes
and sports. The center hosted the parties of the local high schools,
a parents' group, and a social club that showed movies and took
residents on fishing trips and to a Shakespeare festival in Stratford,
Ontario. Adults gathered there for acting lessons, Hawaiian-themed
teas, and for lectures by visiting scholars—the African American

historian John Hope Franklin, an expert on slavery and Reconstruction, spoke to Cabrini residents about the two-century struggle for freedom and equality. When Bobby Kennedy showed up at Cabrini in 1963 as part of his brother's presidential commission on juvenile delinquency, he walked the buildings and shot pool with teens at the Lower North Center. "The feminine part of the staff agreed to keep the hand that the Attorney General shook unwashed," noted the North Side Observer, a regular feature in the *Chicago Defender* that was written in the decorous style of a high-society registrar. Dolores Wilson was present at many of the affairs chronicled in these pages. She helped out with everything from back-to-school giveaways to Christmas bazaars, raffles, and dances. She was a soloist in a "musical extravaganza" and was among the diligent volunteers of a PTA rummage sale to whom the newspaper offered a "low bow" of gratitude for being "on hand into the wee hours of the morning with the 'detested sorting.'"

The Cabrini Extension high-rises were a low-income development of some ten thousand people crammed together on isolated plazas set within a historic slum. There were bound to be issues. Shortly after the Wilsons moved in, drivers delivering milk among the fifteen towers complained that teens were sneaking into their trucks and stealing bottles. In the enclosed stairwells and elevators, hidden from sight, children wrote over the walls. Laundry machines were broken and mailboxes bashed so they couldn't open. In one of the high-rises neighboring the Wilsons, two men on the seventeenth floor robbed another man of his television and then, as a daily reported, "shot him in the eye for no reason." On the corner of Division and Larrabee, on the northwestern edge of the complex, a mass of teenagers brawled.

But Dolores felt like her six-flat on the South Side had experienced more disturbances than the entirety of her nineteen-story high-rise. And at Cabrini, she and her neighbors got together to try to deal with problems. They formed self-protection clubs and

patrolled the laundry rooms, elevators, and corridors. Residents
petitioned the housing authority to enforce rules, respond to maintenance calls, and support programs that engaged the large number
of children living there. They started a law-and-order commission
and demanded that the police carry out foot patrols and do something about the drunks from the taverns on Chicago Avenue and the
roaming prostitutes from the warehouse district to the south. Saul
Alinsky, the father of the Chicago style of community organizing,
came to the auditorium at Saint Philip Benizi school to advise a new
tenant leadership group that called itself A Better Cabrini Organization. Alinsky told the Cabrini residents what they'd already realized
from their experiences: they would have to show power in numbers if
they hoped to exert influence over the CHA, the police, and city hall.

MAYOR RICHARD J. Daley, his son will tell you, never aspired to
have public housing high-rises built in Chicago. "It's a big fallacy that
he wanted all this. It's totally bullshit. He knew more about it than
anyone else. He knew this would make them like a prison." Richard
M. Daley, the mayor of Chicago from 1989 to 2011, has declared this
many times when asked about the legacy of the city's public housing,
his eyes fixed on the pages of a US Senate hearing transcript from
July 1959. The first Mayor Daley had gone to Washington that year to
complain that the federal dollar limit allotted for each unit of public
housing was prohibitively low, allowing only for the construction of
densely packed towers. Daley explained to the senators his hope "to
make it not only high-risers but also walk-up and rowhouses." What
the first Mayor Daley may have desired above all else was a bigger
share of the federal kitty. But his son sees in the Senate hearing proof
that his father has been remembered mistakenly as the creator of
Chicago's public housing system. "'This is not going to go well,' as
my father explained," the younger Daley clarified. "Of course, the
federal government, always in their wisdom, goes ahead with it."

It's true that Richard J. Daley took office after the city council voted to place new developments in existing black neighborhoods, after Elizabeth Wood was fired from the CHA, and after 15,000 of Chicago's 43,000 units of public housing were already built or under construction. At the ceremonial groundbreaking for the Cabrini Extension towers, on April 23, 1955, Mayor Daley dug up several piles of dirt with a silver-painted shovel as photographers snapped his picture. A crowd of a thousand onlookers broke through police lines in an attempt to shake his hand. It was one of Daley's first official acts as mayor, as he'd been sworn in only the previous day.

Born in 1902, Daley grew up in a bungalow in the South Side neighborhood of Bridgeport, a mostly Irish community that included the Union Stockyards, which put forty thousand people to work dismembering and packaging most of the nation's meat. In the nineteenth century the area had actually been called Hardscrabble, and the Irish immigrants who filled it took jobs digging the nearby Illinois and Michigan Canals. Daley, as a teenager, was president of the Hamburg Athletic Club, a group of local toughs, boys from the same parish and operating out of the neighborhood ward office. Daley's "youth organization" was similar to other groups of teen boys found in many impoverished ethnic or migrant communities; they fought as sport, defended their blocks, and usually aged out of street violence as they took on new roles available to them. In Bridgeport, depending on one's abilities, that meant a job either on the "disassembly" lines of the stockyards, in municipal work, or in politics. Daley was a short, pugnacious man whose resting face was a tight-lipped scowl; he possessed neither an abundance of charisma nor eloquence, and his rise in the Cook County Democratic organization was not meteoric. He held a series of positions within the rigid hierarchy of the party, showing fealty to superiors but always lobbying to move up the ranks. He worked in the county treasurer's office and as a state representative, a state senator, the state director of revenue, and as county clerk. An intensely private man robed in tailored suits, he proved hardworking and tough and honest. In 1953, he was named chairman of the Cook

County Democratic Central Committee, second in command behind only Mayor Kennelly. Kennelly, summed up by the journalist A. J. Liebling as "a bit player impersonating a benevolent mayor," had failed to quell racial unrest in several transitioning neighborhoods and made the greater error of cutting back on patronage positions in favor of nonpolitical civil servants. In the next election, the party put Daley's name, not Kennelly's, on the ballot.

The city's famed Democratic machine, which controlled Chicago politics for fifty years, was the invention of Anton Cermak, who took over as mayor in 1931. Cermak would go on to operate it for only a brief time. Two years into his mayoralty, while in Miami, he was greeting president-elect Franklin Roosevelt when an Italian immigrant, inflamed by a general disdain for the rich, fired several shots at FDR. The assassin missed his target but managed to strike five bystanders, including Cermak, who later died from his wounds. "I'm glad it was me instead of you," Cermak supposedly uttered to Roosevelt on the way to the hospital, the words later inscribed on his tomb in the Bohemian National Cemetery, on Chicago's North Side. Chicago under the machine was often called "the city that works." But output and efficiency are at best by-products of machine politics, which is fundamentally a maker of self-perpetuation. A ward boss doles out patronage jobs, and the secretaries, park attendants, and sanitation workers who owe their livelihoods to him raise money for the party, and come election time they get out the vote for the entire slate of machine candidates; the ward, in turn, is rewarded with city services, investment, and additional jobs to dispense, while the party remains in control of the purse strings.

Once in office, Daley wasted no time demonstrating how he would consolidate power over his next twenty-one years as mayor. At his inaugural he announced that he was relieving city council members of their "administrative and technical duties." Constituents had previously gone to their aldermen for everything from requesting a building permit to reporting a road in need of repaving. No longer.

Now they would travel downtown, to city hall, for a brief audience with the mayor in his fifth-floor office. Everyone in the city needed to know that they were indebted specifically to the "Man on Five." "There can be no organizations within the Organization," Daley famously pronounced. In his first two years as mayor, Daley increased the number of patronage jobs by 75 percent, and the number of exams administered for civil servants was cut by more than half. He oversaw a top-down system of leveraged loyalty and reciprocal favors. Daley believed in its virtue without apology. Once, when urging President Lyndon Johnson to appoint one of his Chicago guys as a US attorney, Daley summed up his case by saying, "But more than that, Mr. President, let me say with great honor and pride, he's a precinct captain."

Like other industrial cities along the Great Lakes, Chicago was swooning by the time Daley took office. "A jukebox running down in a deserted bar," Nelson Algren wrote of Chicago in the fifties. The population fell fast from its 1950 peak of 3.6 million. Businesses abandoned the city at about the same rate—Chicago lost more than fifty thousand manufacturing jobs in the seven years before Daley's first term. Daley's plan to "save" the city focused primarily on the central business district and, under his administration, the city would build hundreds of new office towers, including the Prudential Building, McCormick Place, the Civic Center, and the Sears Tower. He presided over the creation of a revised zoning code that freed up where developers could build luxury residential high-rises. And even as people departed for the suburbs, Daley made it easier for them to travel back to the flagging city for business, commerce, and pleasure by constructing highways, the largest underground downtown parking system in the country, with eight thousand spaces, and underground walkways leading directly to office buildings and department stores. You could return to the changing inner city without ever having to be exposed to it.

Daley might have argued for low-rise public housing before the

Senate, but if "high-risers" were what was available, high-risers were what he would build. Millions of federal dollars were at stake, representing thousands of jobs and many big union contracts. As Chicago made upgrades to its centrally located communities, public housing also became a tool wielded to assist the developers involved in the rebuilding. Those displaced by government-funded urban renewal had to be relocated somewhere; they were moved into the new high-rise projects. Almost all these families were African American—not for nothing was urban renewal referred to ruefully as "Negro removal." Whites still made up 13 percent of the families in CHA properties, but they lived almost exclusively in low-rise developments and senior buildings. Daley had inherited a faulty public housing system, but he was responsible for doubling its size. With 43,000 units, Chicago became home to the second-largest stock of public housing in the nation, well behind New York, with 180,000 units (and technically behind Puerto Rico, too).

Most of the new public housing went up on the South Side, with densities greatly exceeding what the CHA had previously allowed. The Robert Taylor Homes, named almost ironically in honor of the agency's African American chairman who'd pushed for integration, replaced a big chunk of the Federal Street Slum and became the largest public housing complex in the world. Its twenty-eight nearly identical sixteen-story towers stretched in groups of three along a narrow ninety-five-acre band. The buildings would extend the State Street Corridor of public housing from the Hilliard Homes, just south of the Loop, through the Harold Ickes Homes, Dearborn Homes, Stateway Gardens, and on to Robert Taylor. It was a largely uninterrupted four-mile wall of public housing, cut off from points west by the newly constructed fourteen-lane Dan Ryan Expressway.

And in 1962, the city completed the William Green Homes, the last portion of what came to be known as Cabrini-Green. The development was named for a longtime leader of the American Federation of Labor, who took over the union from Samuel Gompers. (The

AFL had been found guilty of illegally excluding blacks from the unions building the Cabrini rowhouses.) The Green Homes consisted of 1,096 apartments, in eight exposed concrete towers of fifteen and sixteen stories spaced within a tidy triangle of land formed by Division Street and the point formed to the north by the intersection of two diagonal avenues—Clybourn and Ogden. The buildings' concrete frames and precast concrete sectionals were of the same sandy beige color, and the towers looked like giant computer punch cards. They came to be known as the "Whites," in contrast to the "Reds" across Division Street to the south. Including the rowhouses and its twenty-three high-rises, Cabrini-Green now consisted of 3,600 units of public housing, all of it on a total of just seventy acres of land.

DOLORES WILSON

WHEN HUBERT WILSON was assigned the graveyard janitorial shift for the red Cabrini Extension high-rises, Dolores decided she'd take a job during the day. She worked on and off at the Spiegel catalog company, sending out letters to people who hadn't paid their bills. She had a position as a receptionist for a doctor, as a clerk at a cleaner's, and for a while at the Department of Veterans Affairs employment center. Just as her mother had done on the South Side, Dolores started going door-to-door for the local precinct captain, traveling up and down her high-rise, talking to every tenant, telling them when to vote and whom to vote for. When the ward organization held its big raffle, she sold tickets, coming out of pocket for any of the ones she couldn't foist on others. It was no small expense. "But that's the way it is," Dolores would say. "It isn't what you know; it's who you know." She was right. The precinct captain came to rely on Dolores, and when a bundle of patronage jobs was being divvied up, he raised her name. In 1966, she started at the city's Department of Water Management, just a few blocks east in the Chicago Avenue Pumping Station, across Michigan Avenue from the old Chicago

Water Tower. She was the only black person there, in an office that
handled paychecks for the entire department.

Many nights she'd hustle back to Cabrini-Green from the water
department, cook dinner for the children, help them with homework,
and then rush out again for a meeting at Holy Family or Jenner
Elementary or the Lower North Center. In addition to his janitorial work, Hubert also coached a basketball and a baseball team
that their sons joined. When they lived on the South Side, Hubert
had been in the National Guard, and at Cabrini-Green he started
a drum and bugle corps that he named the Corsairs. Three other
janitors helped him; two were named Brown, and for their different
complexions Dolores called them "Red Brown" and "Black Brown,"
though not to their faces. The men went to Montgomery Ward to get
the fabric to make the Corsairs' pirate uniforms. Neither the Wilson
children nor many of the other boys and girls in the Corsairs could
read music, but they learned the beat and picked up the songs
easily enough. The boys who didn't play instruments turned mock rifles in lockstep, and the girls twirled batons and flags as majorettes.
They practiced in the field outside 1117 N. Cleveland and the other
surrounding towers, people looking down on them from a hundred
different ramps or watching from in front of the buildings. Soon the
Cabrini-Green Corsairs were traveling to suburbs outside Chicago,
marching in parades for Saint Patrick's Day and Memorial Day, and
winning trophies at competitions.

Occasionally, amid their work jobs and their volunteer jobs and
all that parenting entailed, Dolores and Hubert found time to go out
together. Sometimes they just sat outside in the park beside the high-
rises, laughing and talking. It was beautiful out there, Dolores would
say, "Green grass everywhere, flowers, no blacktop or broken glass."
Hubert loved jazz, and he'd invite people over to the apartment to
listen to records. But the two of them also went out with the other
janitors and their wives. South of them, behind Montgomery Ward,
was a ghost town of warehouses come night. But there were some

clubs on Chicago Avenue that they frequented. Dolores favored the Chicago Lounge. It wasn't fancy, but the people there treated them like regulars, remembering their names and what they drank. The DJ spun records Dolores wanted to hear. If someone asked her to dance, she'd have to get Hubert's permission. "He was jealous, jealous, jealous," she would say. "That man wouldn't believe a black cow gave white milk." She didn't give him anything to worry about; she was ever faithful. But if they were apart for ten minutes, he'd ask where she'd been and with whom. They'd argue some nights over these petty exchanges, though come morning they never parted without an apology, trading kisses and saying, "I love you." Dolores had been a teenager when they married, and the two of them still walked the land at Cabrini-Green holding hands.

At the Chicago Lounge, Hubert would look over the guy who wanted a dance, sizing him up, and say, "Yeah, go on, baby." It wasn't close dancing. It was jitterbugging, the bop, the chicken. "But heaven forbid that guy asked me for a second dance," Dolores remembered.

3

Catch-as-Catch-Can

KELVIN CANNON

ON THE WESTERN edge of Cabrini-Green, in the industrial wilds between the river and the high-rises, somewhere within the dumping grounds beneath an elevated roadway, an old witch was said to make her home. Kelvin Cannon had heard the story a thousand times. His mother told it to him at bedtime. His relatives and friends' mothers repeated it. His teachers cited it as a warning. The witch roamed the hills under the Ogden Avenue Bridge, just beyond the Whites of the William Green Homes. Any little boy or girl caught outside after dark, the witch would snatch up. Kelvin had even seen a sketch of her pinned to the bulletin board at the YMCA, like a wanted poster. A gypsy lady, with colorless skin, long, matted hair, and flowing robes of rags.

Kelvin's family was living in 534 W. Division, one of the new white high-rises, when he was born in 1963. His parents had moved there from the West Side, and before that from a river town in Mississippi. Kelvin had convulsions as an infant, and from a young age he believed himself in possession of uncommon powers of perception. He figured the tale of the witch had to be far-fetched. The more he considered it, the more he understood that adults probably put the story out there to keep children like him out of harm's way. The barrens under the Ogden overpass were nowhere safe. You'd find broken-down cars there, machine parts, cracked bits of road, and makeshift tents of cardboard and wood in which vagrants sometimes slept. His mother and teachers could just as well have said a big brown bear lived down there. Yet Kelvin had friends who swore they'd seen the witch. From the upper floors of their high-rises, these children pointed out their windows at the bridge directly below—*Right there! That moving shadow, that's her!* One boy starting his paper route in the predawn gloom was startled when a figure zipped past him. He watched, unable to move, as the white blur slipped through a gap in the fence and disappeared into the hollow below the bridge. Then the boy ran.

Kelvin stayed clear of the bridge after sunset. It was a precaution. He wouldn't even walk above it after dark. But in daylight, that was another story. The Ogden Bridge became his playground. The roadway slashed diagonally across the city's right-angled streets, and at Cabrini it formed into an overpass and veered sharply, nearly brushing against the back of the 1230 N. Burling high-rise. Cars regularly took the curve too fast, and on rainy nights Kelvin could hear from the heights of his apartment the scream of tires followed by the crash. He was sure people died. The next morning, he and other boys would pick through whatever wreckage was left behind. They sometimes played jacks on the bridge's deck. They scampered up and down the stairs leading under it, their dares prompting jumps from ever-higher perches. By the piers below, they gathered up old mattresses and torn-out car seats and flipped off them. He and his friends climbed a mountain of coal. There was a steep hill—Hickory Hill, they called it. The boys took turns riding down it in shopping carts, unable to slow themselves or stop, even as they rolled into traffic. It was thrilling. Sometimes they'd break open the shipping containers stored under the bridge. You could turn them into a clubhouse. A boy who bought a dog training manual kept puppies there. They made slingshots and hunted the pigeons nesting in the rafters. An old woman in apartment 303 of Kelvin's building paid twenty-five cents a bird. Once when a group of them were throwing rocks at the windows of the Old King Coal Company, a police car pulled up alongside them. The cops took hoses from the trunk of their cruiser and wore out their arms beating the boys.

In Kelvin's building the boys hung out together in a great pack, two or three dozen of them. They were children Kelvin's age along with their brothers and cousins. "Every day was another adventure at Cabrini-Green," Kelvin would say. They picked sides for games of football. Or they played sixteen-inch softball or baseball, and they took on the teams of boys from the surrounding high-rises. When they weren't venturing under the bridge, the boys sometimes set out for a game of catch-as-catch-can. They'd start in the morning, each of them darting off in a different direction and trying not to get tagged. They'd sprint through the canyons between the white high-rises, dash up Division, and race across the Ogden Avenue Bridge, over the river and a maze of rail lines and squat factories. They kept on running, over other bridges, through the surrounding parks, along the banks of the river, past the crowds on Chicago Avenue and the shops on Wells Street and back to Cabrini-Green. Ogden Avenue provided a straight shot to the West Side, to the massive Cook County Hospital and into the neighborhoods where Kelvin and many of the other boys had family. But the game's only rules were they couldn't leave the Near North Side. Games sometimes lasted an entire day, the boys' lungs burning, their legs wobbly after five or ten miles. When one of them was caught, the other guys jumped on him a bit, roughing him up with punches and kicks. Kelvin liked that about the game—it toughened them. He felt you needed that. "Growing up in Cabrini, you had to have heart," he'd say. There were hundreds of other rowdy thrill seekers roving about, the kids from each building forming their own packs. Your mom might sew pockets into your underwear so older boys didn't take your money on the way to the store. But Kelvin learned that you at least had to *act* like you weren't afraid. There were kids who wouldn't trade blows, who became targets, who couldn't fight their way to acceptance. Those kids stayed inside. They missed out on the fun, Kelvin said. "Those who weren't scared lived just about a normal life."

Kelvin had heart. Few things scared him. But on his first day of school, at Jenner, he could not stop crying. The elementary school was across Division Street, in the basin below several of the red Cabrini high-rises. Each morning, children spilled out of the nearby towers like coins from a slot machine, all those kids somehow fitting into Jenner, though just barely. It was the most crowded school in Chicago, with more than 2,500 students in a turn-of-the-century building meant to serve half that number. When his teacher, Ms. Redman, tried to console him, Kelvin could tell she was a nice woman. He was able to see people's character like that. She assured him that everything would be fine. But Ms. Redman intimidated him as well. Although Kelvin had sprinted past countless white people on the streets of the Near North Side, he'd never before exchanged two consecutive sentences with one.

IN 1966, MARTIN Luther King Jr. moved into a West Side tenement, as part of his Chicago Freedom Movement. For King, it was an ambitious effort to expand the fight for civil rights from voting rights in the South to fair and open housing in northern cities. The extent and ferocity of Chicago's segregation made the city a fitting target. After he was stoned by angry residents during a march through all-white Marquette Park, on the city's Southwest Side, King said, "The people from Mississippi ought to come to Chicago to learn how to hate." The city's large public housing complexes were the most prominent landmarks of its segregation. By the late sixties, what King called "the Chicago Housing Authority's cement reservations" were home to some 143,000 people, almost all of them black. When King came to Cabrini-Green, speaking at the neighborhood's Wayman AME Church, he showed up in support of residents who wanted changes at the overcrowded and underfunded Jenner. Parents demanded that the school be equipped with science and foreign-language labs, with an adequate library and the remedial tutoring that many of the students required. They insisted that the white principal be fired for her racial bias. A newspaper story about the demonstrations included a photograph of little Cheryl Wilson, Dolores and Hubert's youngest daughter, her pigtails bouncing as she marched with others, chanting that the principal must go.

Three years earlier, when 200,000 black students in Chicago boycotted their segregated and inferior public schools as part of a coordinated "Freedom Day" protest, Mayor Daley responded by saying that in Chicago "there are no ghettoes." But in the city's black neighborhoods, students were being taught in "Willis Wagons," mobile homes named for the school superintendent. The Willis Wagons were parked outside aging buildings, their classrooms filled beyond capacity, while schools in nearby white areas were underenrolled. King returned to Cabrini-Green again in 1966, when Jenner parents were in the third day of a planned five-day walkout of the school. Forty truant officers from the city had flooded the housing development that morning, going door-to-door in the rowhouses and high-rises, warning tenants that they could face criminal charges if they continued to keep their children at home. A rally was supposed to be held in Dolores's high-rise, but the large numbers forced it into Saint Matthew's, on Oak Street. "Should you in any way be persecuted or prosecuted for attempting to seek the best education possible for your children," King said from the pulpit, "I can assure you that thousands of parents from all over the city will come to your aid and together we will join you in jail if necessary." Although the public school superintendent who replaced Willis blamed the protests on outside agitators and a handful of parent provocateurs, the organizing compelled him to transfer Jenner's principal at the end of the school year.

The protests weren't always so peaceful. A few months later, at Waller, one of the neighborhood high schools, hundreds of black students took to the streets, believing that a group of white teens had thrown a black classmate onto the El tracks. The integrated school, located a mile north in tonier Lincoln Park, had recently seen hundreds of white students transfer out. Violence spread to nearby Cooley

High, a predominantly black vocational school that had taken over the building vacated by Washburne Trade. Students there exchanged gunfire with police, smashing the windows of Paul Bunyan's restaurant, Barbara's Bookstore, and other local shops in the nearby Old Town neighborhood. They filled bottles with lighter fluid, set them on fire, and tossed the Molotov cocktails at passing cars. Waller students rallied again soon after, as they demanded that their school offer a class in "Negro history" and hire a black instructor to teach it. Police diverted traffic away from Cabrini-Green for two hours while teenagers threw rocks and bottles to the chant of, "Let's break up Old Town!"

The CHA had celebrated its first developments in the forties and fifties as "Children's Cities," havens for the most vulnerable and blameless of the city's inhabitants. "Chicago," the agency announced, "must plan for those other children who, through the benefits of public housing, can also become the good citizens of tomorrow." The high-rises were designed to accommodate large families: Cabrini-Green's towers included numerous four- and five-bedroom apartments. In just about any neighborhood in America, whether the old Little Hell or Bronzeville on the South Side or the "Bungalow Belt" extending from the northwest corner of Chicago all the way around to the southeast, the population averaged roughly one child for every two adults. Yet in Chicago's family public housing, the ratio skewed to more than two children for every one adult. Of the Robert Taylor Homes' 27,000 residents, nearly 21,000 were minors. In the late sixties, a property manager at Cabrini-Green reported that 20,000 people lived in the complex's 3,600 apartments, 14,000 of them under the age of seventeen.

For the young Kelvin Cannon, that sort of overpopulated children's city provided endless entertainment. "Just to keep ourselves occupied, we strayed into mischief," Kelvin would say. "We had to make up our own fun." A mother might take one child with her to run an errand, leaving the rest unsupervised. They broke into the laundry and storage rooms. They gathered old bedrails or a stripped stove and tossed parts down the stairwells or off the ramps. They leaped up to shatter light bulbs. They had a game in which they pried open the elevators from inside the cabin, propping them with a stick. In the instant that the elevator passed each floor, you could jump on or off. Or you could surf on top of the cabin as it soared the height of the towers. The toll of this mischief was high. Much of the playground equipment around them broke from overuse. The elevators couldn't handle the games, and the CHA was unable to keep pace with needed repairs. The agency said it had to replace eighteen thousand light bulbs every single month across its developments, and it found itself spending a huge portion of its operating budget on elevator maintenance alone. Residents too often were left to climb the stairs in the dark. At Cabrini-Green, the maintenance teams racked up a list of 1,200 unfilled work orders. Leaks, cracked walls, and broken doors went unfixed. The mischief had greater costs as well. Robert Payne was a ten-year-old in one of the red Cabrini high-rises. He leaped from a beam inside his elevator shaft and landed with two feet solidly on the roof of a rising car. His eight-year-old brother, David, jumping next to him, fell short and was crushed between the elevator and the shaft.

President Johnson's War on Poverty identified these "cement reservations" as a key battleground, funding programs in public housing projects with a mission of "maximum feasible participation." That meant Cabrini residents ran initiatives out of the Lower North Center paid for by the federal Office of Economic Opportunity—neighborhood councils, Job Corps, Model Cities, Neighborhood Youth Corps. Tenants took pride in their neighborhood and were concerned about the well-being of the families around them. But the Cabrini-Green that Kelvin Cannon was growing up in was different in fundamental ways from the one Dolores and Hubert had found so uplifting little more than a decade before. When the Wilsons moved to their high-rise, public housing was intended to pay for itself through rent collection. By the time the white William Green towers were built in the sixties, however, the demand for housing in Chicago had diminished. Hundreds of thousands of white residents had already left the city for the suburbs, and the rush of black families arriving from the South had subsided. To fill its 168 high-rises, the CHA accepted tenants with no screening required, some directly off the welfare rolls. Federal reforms to public assistance in the late sixties meant to help the neediest required that local authorities house tenants in order of application, prohibiting the ability to vet for those with jobs. The number of Cabrini-Green households on public aid rose steadily during the sixties to more than half the total population. By the end of the decade, 60 percent of the families with children had only one parent at home. Women on welfare were in many ways discouraged from marrying or officially sharing a residence with the father of their children, since the presence of a man could leave them ineligible for benefits. The median income of CHA residents dropped from 64 percent of the citywide average in 1950 to just 37 percent in 1970.

Public housing residents in Chicago paid different amounts in rent depending on their incomes. In 1960, rent for a three-bedroom started at $41 a month and went up $1 for every $55 more a tenant earned annually, up to a maximum rent of $110. As the residents as a whole became poorer, the amount the CHA received from rents plummeted. Federal funding, rather than making up the difference, was scaled back as well. Across the country, housing agencies ran huge deficits. For several years starting in the early sixties, the CHA switched to fixed rents—$80 for a three-bedroom, or $140 for the same unit if a family earned above a certain threshold. It was simpler, and did away with tenants hiding income, but it was harder on the poorest families on welfare. Edward Brooke, a Massachusetts Republican and the first African American ever to be popularly elected to the US Senate, sponsored a bill that he hoped would protect those in public housing from shouldering the burden as housing authorities scrambled to get out of the red. The 1969 Brooke Amendment pro-rated rents at no more than 25 percent of a household's income (eventually upped to 30 percent). The law, while well meaning, meant a big jump in what working families who'd been paying a fixed amount each month now owed. Already putting up with broken elevators and inferior schools, this stabilizing force now had another reason to abandon public housing, and their departure left behind an ever-greater share of residents who were on public aid and paid very little in rent. "These rental increases are intolerable. Families with jobs are being forced to move out," the president of the Cabrini-Green tenant council wrote in a letter sent to city, state, and federal officials. "The small savings to the U.S. Treasury must be balanced against the devastating impact on the lives of many public housing residents, especially when other benefits for the poor are being cut."

CHA properties required constant upkeep—tuck-pointing and weatherproofing, repairs to roofs and masonry, to windowsills, electrical systems, plumbing, heating, and appliances. With less money coming in, the CHA cut back on maintenance. To save money at Cabrini-Green and other developments, the landscaped perimeters were paved over. "Everything became blacktop," Dolores Wilson said. "No flowers. No trees. No nothing." After 1968, the ramps were also fenced in from floor to ceiling at almost all the high-rise developments. A precaution to protect both residents from falls and people below from falling objects, it also made tenants appear as if they were caged inside their buildings. With the concrete boundaries and the metal fencing, the high-rise public housing complexes took on the look of a prison.

The CHA's mismanagement of its properties was also monumental. No other housing authority in the nation spent more on labor costs, and no city got less for its money. The head of the agency beginning in the sixties was a consummate city operator who regularly traded on his position in government to enrich himself. Charles Swibel was ten when his family emigrated from Poland, in 1937, and he spoke no

English. He spent his afternoons in Chicago educating himself in his neighborhood public library, and at fourteen he won a citywide essay contest on the patriotic theme of "What America Means to Me." He took a job filling mustard packets at a kosher sausage factory, and another with one of the city's largest slumlords, sweeping the offices but soon graduating to rent collector in the rooming houses along West Madison Street's "Skid Row." The owner of these down-market buildings paid Swibel's way through college and eventually promoted him to president of the company. (The man's children would later sue Swibel for raiding the family trust fund, with Swibel settling the case out of court.) Mayor Daley appointed him to the CHA board in 1956, and Swibel took over as chairman of the agency in 1963. Under his leadership, the CHA suffered from a bloated staff and fake work assignments, and Swibel, whom the daily press liked to call "Flophouse Charlie," awarded contracts to cronies. Time and again, he ignored calls for reforms after scathing reports documented his agency's failings.

Swibel also borrowed money at extremely favorable rates from those doing business with the CHA, using the funds to finance his private development projects. One of them was Marina City, the circular twin towers bordering the Chicago River north of the Loop and a mile from Cabrini-Green. As an appeal to young white professionals nervous about urban living, the buildings were marketed as a protected "city within a city," with their own theater, bowling alley, restaurants, and boat slips. The corncob high-rises were also key to a burgeoning revitalization of the Near North Side in the mid-sixties that both skipped over and further isolated Cabrini-Green. Chicago's white population declined by more than 800,000 from 1950 to 1970, and the river district that included Cabrini-Green flipped from 85 percent white in 1940 to 90 percent black thirty years later, forming the only significant African American settlement in the city not on the South or West Sides. Yet even as the housing project seemed to expand to define its immediate surroundings, the city was already taking action to "rescue" the nearby communities from a similar fate. Nearly a hundred acres in this area close to the city center were declared an urban renewal site; three-quarters of the existing housing was razed, and the emptied land readied for development. In 1962, the Chicago real estate baron Arthur Rubloff broke ground on a sprawling middle-income residential complex to the east of Cabrini-Green, in the Old Town neighborhood, using land bought and cleared by the city's Department of Urban Renewal. The 2,600-unit Carl Sandburg Village included tennis courts, swimming pools, and underground parking. Its nine residential towers were each named for renowned literary figures: Cummings House, Dickinson House, Faulkner House. Many of the new residents were young and white and working in the Loop, and the Puerto Ricans who'd lived on the cleared land were pushed farther north. If it had not been for Sandburg Village, Rubloff claimed, "the whole area would have gone down the drain."

Cabrini-Green was contained in other ways as well. North Avenue, the boulevard separating the housing project from Lincoln Park, was widened, extending the gulf. On the Lincoln Park side, facades were remade so that entrances opened onto opposite-end courtyards, and roads were turned into cul-de-sacs. In the sixties, a new group called the Lincoln Park Conservation Association, citing the damaging effects of blight on its members' community, petitioned the city to disconnect Ogden Avenue after its bridge passed north of the white high-rises of Cabrini-Green. Since the 1930s, the slanting thoroughfare, named for Chicago's first mayor, had offered West Siders a speedy route to the beaches and shops along the lakefront. But Lincoln Park residents now said the wide thoroughfare was an eyesore and harmful to development. In 1967, they convinced the city to close the street to traffic at 1230 N. Burling, the barricades there forming a literal and symbolic divide between Cabrini-Green and the gentrifying areas north of North Avenue. Witch or no witch, Kelvin and his friends could have at their junkyard games beneath the dead-end bridge. But a couple of blocks past their housing project, Ogden Avenue disappeared under newly built parks, townhomes, and a pedestrian mall.

IN APRIL 1968, Kelvin was just tall enough to stand on tiptoe and peer over the windowsill in his seventh-floor apartment. He watched wide eyed as people broke into stores and pushed overflowing shopping carts down the middle of the street. The day after Martin Luther King Jr. was assassinated, students at the schools around Cabrini-Green rushed out of their classes; teachers locked themselves inside the buildings, waiting for the police to escort them to their cars in dashes of three and four at a time. People who filled the streets pulled over delivery trucks, beating the drivers and cleaning out the vehicles of their goods. Looters tore through the local stores— Jerry's, Del Farm, Pioneer, the A&P on Clybourn, Greenman's, Harry's drugstore, Big Frank's.

Dolores Wilson watched from her high-rise as well. She saw men raid the cleaner's and walk out with pants and shirts and jackets that probably belonged to people they knew. One lady left a butcher shop with a whole hog. "Where will she put all that meat?" Dolores asked herself. "When will she eat it all?" There were people with carts full of greens. The Wilsons didn't have very much to eat at home, and Dolores thought she could use some of those greens. Someone might bring them at least an egg or two, she joked. Kenny was twelve, and he begged his mother to let him go outside to join his friends. "You don't do what everybody else does," she scolded him. "You do what I say." Che Che was older, however, a teenager, and he was on his way to a pool hall when members of the National Guard arrived. He ducked behind a car. "Are you one of them looting?" a cop in riot gear asked him. "No, sir." But they clubbed him over the head anyway and took him to lockup. When he didn't show up at home that night, the Wilsons called the jail. They were told no one named Hubert Wilson

Jr. was there. But they had a friend who worked at the jail, and he eventually found Che Che's paperwork. Their oldest son walked out with a bandage wrapped like a turban around his head.

Several blocks in any direction from Cabrini-Green, business went on as usual, and the riots were mostly experienced as a news event on TV. But at Cabrini-Green, the cracks of rifles were heard for days. Shops on North, Larrabee, Oak, and Wells were ransacked. Some residents were too fearful to venture outside, and local agencies brought them food and other essentials. Crowds continued to mill about the streets, primed to vent their frustrations. Shouts went up that they should march over to Sandburg Village. The new private housing development seemed to represent everything that Cabrini-Green was not—and also what Cabrini might become, since Sandburg's developer, Arthur Rubloff, said publicly that he could convert the Cabrini towers into condos and make a fortune. But Cabrini residents didn't make it over to Sandburg Village. They heard about Mayor Daley's orders to "shoot to kill." And National Guardsmen in tanks positioned themselves on Wells Street, the eastern boundary between Cabrini-Green and the Gold Coast beyond; police sealed off the still-existing westward portion of the Ogden Avenue Bridge. Those at Cabrini were hemmed in on all sides.

On the West Side, in predominantly black neighborhoods, people burned stores, as well as the apartments above them. More than two hundred properties were destroyed. By comparison, the physical damage around Cabrini-Green seemed manageable. Life went on there, if uneasily. Some of the longtime shop owners chose not to restock their looted establishments. Germans, Italians, and Jews who'd already witnessed a couple of the demographic transformations on the Near North Side decided that they no longer belonged. But most of the stores reopened. Cabrini residents boycotted a few of them for price gouging. The Black Panthers set up a free breakfast program in the basement of one of the Catholic churches left empty by the long-gone Italians. Students from Waller and Cooley went ahead with plays, choruses, and formals. Retirees met for luncheons at the Lower North Center, and children enjoyed an Easter egg hunt with games and songs in the nearby 135-acre Lincoln Park. "The curfew had taken its toll," the *Defender* reported of the Easter event, "and the youngsters were more than pleased to get into the routine of evening Recreation."

KELVIN CANNON

NOT TOO LONG after the riots, Kelvin Cannon's parents split up. He would blame his mother for pushing the quarrels nearly to the point of violence. But it was his father who left for the South Side, where he soon remarried and started a second family. Kelvin's aunt, who lived in another white Cabrini-Green high-rise, 714 W. Division, offered to help her sister with the children. When Kelvin was seven, in 1971, the family moved two buildings down, and his mother went on welfare.

Kelvin instantly made two best friends in his new building. He practically lived in Reginald and William Blackmon's apartment, and they in his. They ate at one another's kitchen tables, slept side by side in crowded bedrooms, and each of their mothers put Band-Aids or spankings on any of them as if they were her own. Like Kelvin, almost all the other children in 714 W. Division had only a mother at home; Reggie and William's older brother, Richard, once counted just five grown men living in the entire sixteen-story, 134-unit tower—which was one of the reasons the boys traded stories about their gym teacher. After the move, Kelvin would no longer attend Jenner; children in his new high-rise crossed the blacktop outside their building and walked to Friedrich von Schiller. The PE teacher there, Kelvin was told in advance, was a professional athlete and would teach you how to play basketball and baseball just like him. His name was Mr. White, and he would buy you sneakers. He would unbuckle his belt in the middle of class and whip your behind. He would take you horseback riding, swimming, and canoeing.

Jesse White really did these things. He coached every sport at the nearby YMCA, and he led the largest Boy Scout troop in the nation out of Cabrini-Green, with over three hundred members. Since 1959, he'd operated the Jesse White Tumblers, a high-flying gymnastics program. Cabrini children did leaping somersaults and acrobatic dives over rows of their fellow Tumblers at everything from Chicago Bulls games to state fairs and broadcasts of *Bozo's Circus*. White once transformed the blacktop around Kelvin's building into a roller rink. Another time, he and a white retired naval officer from Sandburg Village named Joe Owen created the Cabrini-Green Community Sandlot Tennis Club; they painted the lines for the courts right onto the concrete, dug holes, planted posts, and set up the nets. They held contests to see which kids could clear the courts of the most pieces of glass—one winner bagged eight thousand shards. Some children told Kelvin that they spent close to 365 days a year with Mr. White, seeing him more than they did their own families.

The first time Kelvin entered Schiller, he expected to see a giant of a man, John Henry in gym shorts. He imagined that the ground would rumble beneath the PE teacher wherever he stepped. But although an ordinary five foot eight, Jesse White cut an imposing figure, with a neck like a stone column and barrels for arms. It was also the way he stood in front of the PE class, his shirt and pants pressed, his back straight, how he instructed the students to assume a military position as they recited the Pledge of Allegiance—chins up, heels touching, and feet pointing out at 45 degrees. He pronounced his words loudly and powerfully but without displeasure, repeating a collection of tough-love maxims. "Winning is for you, losing is for someone else." "Be a gentleman first, we'll argue about the rest later." He had a dozen of these sayings that seemed from a different place and time, more *Knute Rockne, All American* than Cabrini-Green. "I want you to be leafless, smokeless—the only time you should use pharmaceuticals is if you're wearing a lab coat." He talked to his students about geography and the environment. He taught his Tumblers how to speak to donors, never to beg for anything. He aimed to keep them too busy to get mixed up in trouble, but he'd also say he was preparing them in case they made it out of the projects. Many of the boys under his charge thought, *I want to be just like him.*

Jesse White was born in 1934, in Alton, Illinois, a city along the Mississippi River, in the southern part of the state. At seven, he moved with his family into a cold-water flat on Division Street, on the very spot where the William Green Homes would be built. An inveterate extrovert, White reveled in the diversity of the Near North Side of his youth. He'd sample Greek and Persian cooking; he'd thrill Italian or Hungarian shop owners by speaking a few sentences in their native tongue. His father worked a dozen odd jobs as he struggled to find something steady, and the family subsisted on public aid for several years. White learned firsthand that welfare was not only a necessity but also a woefully insufficient one, incapable of adequately feeding and clothing a family. But he believed that one could make the choice to transcend poverty. It might require the generosity of others, but your own rigor and rectitude were essential. As a teen, he went door-to-door in the tenements delivering coal, ice, and wood. He could also jump three feet when he shot a basketball at Waller High, and he was the first person in his family to go to college, using a basketball scholarship to attend Alabama State University, a historically black school in Montgomery. In 1955, he'd say, he introduced the jump shot to the heart of Dixie. He was a student during the Montgomery bus boycott; the civil rights leader Ralph Abernathy pledged him into the Kappa Alpha Psi fraternity; Martin Luther King Jr. was the preacher at his church. White wouldn't sit in the back of a bus, but neither did he join the protests. And he never lost faith in the institutions of the country or in his own personal ability to overcome.

White was far from the only person at Cabrini-Green who hustled and scraped and labored to keep the children there productively busy. Elax Taylor operated the 911 Teen Club out of the basement of the 911 N. Hudson high-rise. Marion Stamps found money from the

federal Model Cities and other War on Poverty programs to open a
health clinic, a youth center, and an alternative school with an Afro-
centric curriculum. Al Carter started a youth foundation, a Little
League team, and the Cabrini-Green Olympics. Hubert Wilson had
the Corsairs. And numerous other residents volunteered their time,
coaching and mentoring, or worked low-paying jobs in nonprofit day
care and after-school programs. But Jesse White had a unique ability to bridge the divide between the children in the towers and the
power brokers close by. He was drafted into the army after college,
in 1957, serving in the 101st Airborne Division. He played in the
Chicago Cubs farm system for eight summers, each fall returning
to Chicago to run youth programs for the park district at Cabrini.
He landed the park district job the same way all city jobs were allo-
cated back then—through the Democratic Party's regular political
machine. White's parents had brought voters out to the polls each
election, and his precinct captain in the ward organization recommended him. White became close with the ward committeeman,
George Dunne, a lifetime Near North Sider who served in the Illinois House of Representatives and went on to become president of
the Cook County Board of Commissioners, the second-most powerful
position in the city behind the mayor. White believed in the bonds
of loyalty that held together the machine's patronage system—you
were elevated so that you could elevate others. Through the party,
he found businessmen who were willing to help him buy vans for
transporting his tumbling team and winter coats and shoes for other
children at Cabrini-Green.

Kelvin was never a Jesse White Tumbler. But he was a Cub Scout
and Boy Scout under Mr. White. As one of Mr. White's Patrol Boys,
he put on a uniform to direct other children to and from school. And
when Kelvin was eight, he was one of fifty Schiller students White
took skiing; none of them had ever been on a slope or worn skis, and
it was the most exciting day so far in Kelvin's life. "He was that father
figure who wasn't at home for a lot of us," Kelvin would say. "He took

us places like a normal father might take us. He spent time with us like we were his kids."

Kelvin had his first real fight soon after starting at Schiller, when he was eight. Kelvin and Reginald Blackmon were playing king of the mountain on a landfill beside a construction site. Two boys from the next-door tower, 1230 N. Burling, came over to challenge them. One of them was Perry Browley, a star of the Jesse White Tumblers who would later break his neck doing a flying somersault when his hand slipped off the satin of his pants and he untucked too soon; everyone thought he'd be paralyzed for life, but Perry recovered and even went on to start his own tumbling program, teaching gymnastics and conflict resolution on the West Side. Back then, though, he was eight, and Kelvin slung him roughly off the dirt heap. Kelvin raised his arms to signal his dominion. But the boy with Perry was two years older. He knocked out one of Kelvin's teeth. Kelvin looked around for help from his friend, but Reggie was already in the wind.

4

Warriors

DOLORES WILSON

"It was real nice and quiet and peaceful in Cabrini," Dolores
Wilson would say about the years before the King riots. "We didn't
have to worry about gangbanging. After Daley put the Stones in Cabrini, that's when we started to see gang writing on the walls. That's
when shots started ringing out. That's how gangs with a plural began
to form at Cabrini-Green." The Stones were the Cobra Stones, a West
Side gang. Fires in West Side neighborhoods after King's assassination left more than a thousand people homeless. The city happened to
be by far the largest property owner in Chicago, and it had vacancies.
In a simple act of deploying resources at a time of crisis, the city
relocated West Side families into 1150–1160 N. Sedgwick, connected
towers at Cabrini-Green just to the north of Dolores Wilson's high-
rise. That's what the social safety net was supposed to do. It caught
people who were falling, saving them before they crashed into the
depths. But it was one thing to operate a housing program and another to run an emergency shelter. At the very least, a sudden influx of
homeless families required a boost in managers and services meant
to assist them. Yet none of those things accompanied them to Cabrini-
Green.

The Cobra Stones were soon operating out of 1150–1160 N.
Sedgwick, and the building came to be known as "the Rock." The

black gangs in Chicago were split into two alliances, the People and the Folks. The People included the Blackstone Rangers, the Cobra Stones and Mickey Cobras, and the Vice Lords; the Folks were the Gangster Disciples and the Black Disciples, both started on the South Side. Because of Cabrini-Green's isolation from other black neighborhoods, the street gangs there were mostly unaffiliated with either of these alliances. But the presence of the Cobra Stones helped change that. Rifles were fired from the windows of the Cobra Stones' high-rise, and teenagers from the towers around it armed themselves as well. If Stones were shooting at them, they were going to shoot back. In one of the neighboring buildings, guys formed a gang called the Blacks, and the blacktop between the close-cropped towers, along the old Death Corner, was called the Killing Field for all the violence it saw.

Standing in this open field, looking up at the walls of cross-hatched red high-rises on all sides as if from the bottom of a ravine, you realized suddenly how vulnerable you were. What before were just buildings staring indifferently at one another now could seem vicious. A thousand windows rising up, and in any one of them a heedless seventeen-year-old with a rifle could smudge you out as he would an ant. A woman pushing a baby in a stroller was almost shot as she passed below. Hubert Wilson's Corsairs stopped practicing on the blacktop. The priest running the nearby Saint Joseph parish opened up his gymnasium for them, but on their way home at night, when tensions between the buildings were high, the boys and girls sprinted across the exposed concrete tract.

Dolores never had any trouble with her daughters, and Che Che was a bookworm. He'd sit out nights on the ramp, under the overhead light, and read past midnight. Her other sons were more rebellious, Michael in particular, who received the greatest share of Hubert's whippings. Michael was among the teens in their high-rise who formed a gang called the Deuces. Dolores wasn't happy about it, but she understood. "If they came out the building on one side, there are the Stones shooting at them," she explained. "They come out the other, and there are the Blacks."

One night, the police grabbed Michael from in front of his building. They took him on the expressway, driving six miles south, to Bridgeport, the Irish enclave where Mayor Daley and many city cops lived. Michael believed that the police were arming the Cobra Stones at Cabrini-Green. He said he'd seen a patrol car pull up in front of 1150–1160 N. Sedgwick, the cops popping their trunk and handing out rifles to the gang members. Now the police dropped him off in Bridgeport. "They didn't even have black cars in Bridgeport," Dolores would say. She and Hubert were sleeping when the telephone rang. Michael was calling from a phone booth outside a tavern. Hubert told him not to move, and to try not to be seen, and he got his son and brought him home. Another time Michael was taken to the police station. When Dolores arrived, a sergeant was asking her son to identify the police officers who had tossed him against a wall. Dolores could see the cops waiting eagerly for Michael's response. The police both ignored and brutalized the all-black population at Cabrini, and she felt sure they were looking for a reason to take him into a back room and teach him what happens to boys from the projects who challenge the authority of the cops. "Shut up, Michael," she shouted. "Didn't I tell you about lying?" Stunned at first, he caught on. Michael said he didn't know who messed him up, and he left with his mother.

With the increase in violence at Cabrini, Hubert bought Dolores a gun. It was a cute little derringer with a pearl handle that she kept in her purse. She was more afraid of the gun going off when she reached for her carfare than any situation in which she might feel the need to use it. Dolores had stayed loyal to her South Side hairdresser all these years, and every few weeks she traveled back to her old neighborhood by bus and train. On a return trip from the South Side, she was on the Division Street bus when three guys got on at Clark, several blocks shy of Cabrini-Green. They surrounded a Mexican man seated two rows in front of her and stuck a nail file to his throat. Dolores started tapping the young men on the back. "Why don't you leave that man alone? This isn't right," she cried. No way was she going to pull out the gun. "I wasn't that much of a good Samaritan," she'd say. "A tap on the back was about all I had in me." When the bus driver passed two police officers, he tooted his horn, and the muggers jumped off the bus and ran into the nearest Cabrini high-rise. The Mexican man reported the crime to the officers, and Dolores, being nosey, got off the bus to listen. But then she felt a sprinkle of rain. With her hair freshly done, she jogged home across the blacktop, a hand covering her head. The elevator doors at her building opened, and she saw a crisp $10 bill on the floor, folded in half. She told Hubert about the entire incident, saying, "See, God knows a good deed."

IN 1969, DAVID Barksdale, the founder of the Black Disciples, merged his gang with the Gangster Disciples, creating the Black Gangster Disciples Nation. The union put an end to street fights between their groups and allowed them better to compete with Jeff Fort's Blackstone Rangers for control of the South Side drug and gambling trade. While gun battles between the Rangers and the Black Gangster Disciples continued, the gangs also drew attention for announcing their shift into legal activities. The "street organizations" said they were now going to be forces for good, using their administrative know-how and manpower to rebuild. The Woodlawn Organization, a grassroots community group that formed on the South Side to fight urban renewal plans in nearby Hyde Park, received almost a million dollars in federal grants from President Johnson's Office of Economic Opportunity to operate job-training programs led by the Disciples and the Rangers. The two gangs ran their own training facilities targeting delinquent teens. Fort's Rangers started a newspaper, a legal defense project, and the Black P. Stone Youth Center.

In North Lawndale, on the West Side, a gang called the Conservative Vice Lords went even further. They launched a street academy for high school dropouts, an art studio, and a neighborhood beautification effort ("grass where there was glass"), as well as employment and housing services. The group opened the Teen Town ice cream parlor, two recreation centers, and the African Lion clothing store. The gang legally incorporated, becoming CVL Inc., and received funding from the Rockefeller and Ford Foundations, from state and federal government, from Sears, Western Electric, and Illinois Bell Telephone, and from Sammy Davis Jr. Members held open houses for the police and met with city officials to prove the authenticity of their about-face. "The turnaround came because we were going to start doing things in our community for ourselves," the CVL spokesperson Bobby Gore said. For a time in 1969, the three major black gangs formed an alliance they called LSD—for Lords, Stones, Disciples—and in a series of demonstrations targeted the city's trade unions for failing to hire, train, and promote African Americans. They shut down construction sites and marched on city hall, the sight of these young black men in berets and sunglasses massed in the Loop a source of pride or terror, depending on your perspective.

The model for going straight wasn't another black gang but Old Man Daley. In the summer of 1919, when Daley was seventeen and president of his Hamburg Athletic Club, Chicago experienced its worst race riot. The violence was triggered when a black teenager at a Lake Michigan beach drifted across the invisible barrier dividing segregated waters; whites responded by pelting him with rocks until he drowned. The larger cause of the rioting was the surge of African Americans then arriving in the city from the South, the first wave of the Great Migration. In neighborhoods like Daley's Bridgeport, on the border of the Black Belt, white residents were ready to push back against the tide. Already that year there had been numerous cases of African Americans moving into these borderlands only to have their homes dynamited. Over the riot's five days, thirty-eight people were killed and hundreds more seriously injured, African Americans making up two-thirds of the casualties. An investigation into the riots found that the Hamburg youth organization was guilty of instigating attacks on blacks, though Daley was never personally linked to any of the assaults. He was able to age out of the group and move on because teens in Bridgeport had jobs awaiting them. If there were comparable opportunities in Chicago's black neighborhoods, CVL leaders argued, then Vice Lords, Disciples, and Cobra Stones likely could also graduate to legitimate careers.

Some of the community work that gangs did was genuine; a great deal of it not. The same could be said of work undertaken by city agencies. But Mayor Daley dismissed it all, declaring any claims by gangs of legitimacy pure subterfuge, their efforts fronts for illicit operations. In a Democratic Chicago without an opposing party, Daley had hardened into a law-and-order conservative. He'd handled Martin Luther King Jr. as a subversive, declared war on rioters after his assassination, and met demonstrations in the city during the 1968 Democratic National Convention with brute force. Daley now announced a new policy that would treat street gangs not as wayward youth but as organized crime. The Chicago police established a special gang intelligence unit, and Daley pressured foundations to cut off funding to the groups, vetoed job-training grants, and prohibited police from holding discussions with any suspected gang members. The gang unit raided the programs that the Disciples and Rangers operated with the Woodlawn Organization. And on an early morning in December 1969, fourteen Chicago police officers stormed a West Side apartment where members of the Black Panthers were living. The officers killed both Fred Hampton, the twenty-one-year-old leader of the Illinois chapter of the Panthers, and Mark Clark, on security detail for the group, claiming self-defense amid a fierce gun battle. Ballistic reports revealed that out of the nearly hundred shots fired, all but one had come from the guns of the police, and Hampton and others were still in their beds. That same month Chicago police arrested several gang leaders on a range of charges, some of which were soon dismissed and others of which stuck. Bobby Gore, the CVL spokesman, was charged with a murder that occurred outside an Ogden Avenue bar and was sentenced to twenty-five to forty years. In prison he was called "Kissinger" for his efforts to keep the peace between inmates, and when he was released, in 1979, he went to work at the Cabrini-Green offices of a foundation that helped other former inmates find jobs.

Within the Chicago Police Department, a force of twelve thousand officers, efforts were also made to repair relationships with the city's black communities. In the fall of 1968, after police officers chased stone-throwing teens into a Cabrini high-rise and maced a one-year-old in his aunt's arms, the department sent forty-five black officers into the housing development to calm residents and avert another riot. The police started a special walk-and-talk beat at Cabrini-Green as well, officers on foot socializing with people, with the hope that cops and Cabrini residents could get to know and even respect one another. Sergeant James Severin and Officer Anthony Rizzato volunteered for the community policing assignment. Severin, thirty-eight, was a former insurance investigator and army corporal; a thirteen-year veteran of the police force, he had worked the security detail protecting King when the civil rights leader moved to the West Side. Rizzato was thirty-five, married, with two children. He and his brother had signed up together to become police officers four years earlier. Both Rizzato and Severin were even-tempered and unassuming, and amenable to the idea that it might be more prudent to throw a baseball with a child in public housing than to arrest him for throwing bottles at cars.

It was still light out on July 17, 1970, at 7:00 p.m., when Rizzato and Severin were on their foot patrol. They walked onto the baseball diamond at Seward Park, below the surrounding red high-rises. Rifle shots came simultaneously from two of the towers overhead, 1150 N. Sedgwick and 1117 N. Cleveland, the Rock and Dolores Wilson's building. Severin and Rizzato were both hit. The officers arriving on the scene couldn't get to them; snipers continued to fire down from the high-rises, and they were forced to take cover. More police arrived, a hundred of them, blasting out windows and screens as they traded shots with the unseen enemy. Military veterans on the force said it was the closest thing they'd experienced in Chicago to their tours overseas. Officers brought in heavy weapons artillery while a police helicopter hovered over the park, shining its spotlight. Patrolmen drove their squad cars onto the infield and used them as shields as they retrieved their colleagues. Severin and Rizzato were taken to a nearby hospital, where they were pronounced dead on arrival.

Kelvin Cannon was six and saw the helicopter whirling outside his window. The police stormed all the buildings and blocked exits. One of Dolores Wilson's neighbors asked the police what was going on, and the cops knocked him down and beat him with batons. The police searched apartments, using battering rams and sledgehammers to break down any doors not opened to them. Officers arrested several residents that night, including anyone who didn't cooperate. Among them was twenty-three-year-old George Knights, a janitor's helper in the high-rises. When a warrant was issued for another suspect, Johnny Veal, a seventeen-year-old Cobra Stone, Veal feared that officers would kill him if he was brought in for interrogation, so with his lawyer he later turned himself in to a judge. Prosecutors said that Knights and Veal were rival gang members teaming up to kill cops. They were each found guilty and sentenced to terms of 100 to 199 years in prison.

As difficult as it was to shock the city after the turmoil of recent years, the cop killings represented for many Chicago's collapse into chaos. On the streets, any regard for authority was lost. "A complete breakdown of law and order in this town when the citizens murder two policemen whose assignment was to improve relations between the police department and the citizens," said a representative of the police. An editorial in the *Defender* called the murders "one of the most revolting of the current epidemic of terrorist acts in the black community." Public housing had been viewed with distrust from its outset, the critiques usually colored by a general wariness of any assistance for the poor. In the Cold War hysteria of the 1950s, residents in the Cabrini rowhouses were forced to take loyalty oaths in exchange for their government aid. But now public housing was also seen as the place where too many welfare-dependent black people were stacked atop one another in too little space. The government housing intended to clear out the slums a mere generation ago was now the "projects" and epitomized all the unruliness and otherness of the city in decline. And Cabrini-Green at this moment became the proxy for all troubled public housing. "That's when it started," Dolores Wilson said. "Now they could say, 'It happened at Cabrini.'"

Just like the Italian slum that preceded it, Cabrini-Green's location only blocks from the Gold Coast added to its emerging infamy. Twelve other police officers had been killed in the previous eighteen months in an increasingly violent Chicago, with eleven of those murders occurring on the South Side. In Englewood, an officer was sitting in his squad car filling out paperwork when someone walked up and shot him point-blank. By 1970, however, most white Chicagoans rarely spent time in South Side neighborhoods like Englewood, and they probably passed the Robert Taylor Homes only in a car speeding along the Dan Ryan Expressway. But Cabrini-Green remained that uncommon frontier where white Chicagoans still crossed paths with poor blacks. They drove or even walked by Cabrini-Green with some regularity. People from Lincoln Park, Old Town, and the river districts by the Loop regarded the housing project as a place to avoid, or to go to for felonious offerings. They followed the reports in the media of late-night violent crime at Cabrini as if hearing about a near miss.

The city's news teams, too, needed only to look in their own backyard to cover another grim story on public housing life. In the days and weeks after Severin and Rizzato were murdered, every incident at or even near Cabrini-Green received extensive news coverage, with journalists exercising their more noirish proclivities, doing their best Nelson Algren. The on-and-off gunfire became "a special kind of rain." The thousands of children at Cabrini were "clenched in a cold

concrete-and-steel fist of seventy acres." A *Chicago Tribune* editorial titled "Near North Hell," reviving the old Little Hell moniker,
described the development as "a Hydra's head of problems and, like
the mythical monster, as one problem is cut away, two more come to
take its place." The *Chicago Sun-Times* ran a special series in which
it sent a black staff writer—actually, a college student, an intern, who
was promoted for the assignment—to live for a time at Cabrini-Green,
the newspaper publishing his dispatches as if they were being filed
from the jungles of South Vietnam. "For several days, he lived there,
talked with residents, roamed the dingy corridors and littered playground, played ball with its teeming young."

Thus Cabrini-Green after the cop killings became both a real
place with an array of actual problems and an abstraction. In short
order Cabrini-Green was the setting for the Norman Lear prime-time
sitcom *Good Times*, which told the story of a family of five struggling
nobly in their two-bedroom, one-bathroom public housing apartment.
James Evans, the family patriarch, orders his youngest son to his room
after he's suspended from middle school for calling George Washington a slaveholding racist. "But, Daddy, this is my room. I'm the little
fellow who sleeps on the couch." *Good Times* was one of the most-
watched programs on television, behind *All in the Family*, *Sanford
and Son*, *Chico and the Man*, and *The Jeffersons*—all of them playing
to the country's fixation on issues of race and class and upheaval in
inner cities. Eric Monte, one of the creators of *Good Times*, grew up in
the Cabrini rowhouses, and the movie he wrote based on his teenage
years there, *Cooley High*, was marketed as the black *American Graffiti*. For the 1970s urban crime novel *The Horror of Cabrini-Green*,
the cover copy promised, "Inside Chicago's worst public housing
project and a young man's vicious struggle to survive." Each of these
representations of Cabrini-Green added to the myth and also came to
shape the reality there. During one brutal stretch of pages in *The
Horror of Cabrini-Green*, an eight-year-old is raped, the police shoot
a child who stole a bag of potato chips, and the sixteen-year-old

narrator, Bosco, and other members of his gang break into a neighborhood church and hang the priest. "'Damn, it sho is a violent day,
ain't it?'" one of Bosco's friends observes nonchalantly, suggesting
they might head over to another high-rise to check out the pummeling of a boy they know. "Shorty and I agreed that it was definitely a
violent day," Bosco relates. "We went over to watch Jessie get his ass
whipped."

In a public housing newsletter, a media-relations official at the
CHA addressed the outsize place Cabrini-Green came to hold in
the civic imagination: "'Cabrini-Green' now stood for something more
than a public housing development in Chicago named for a Catholic
saint and a labor leader: it also spelled 'fear' in the minds of the citizens of the city and the residents of 'the project.'" Defining Cabrini-
Green as *the* big civic problem also meant it couldn't be ignored; it
needed to be dealt with, solved. "As though by focusing on this one
community and isolating it," the CHA official said of the sudden
obsession with all things Cabrini, "they could exorcise the violence,
the crime, the poverty, and the racial fears of the city."

CABRINI-GREEN'S NEWFOUND NOTORIETY brought Jesse Jackson
out to the housing project. In 1970, Jackson was running the Chicago chapter of Operation Breadbasket, the branch of the Southern
Christian Leadership Conference focused on improving economic
conditions in African American communities. At a news conference
in front of the high-rises, he praised Severin and Rizzato while evoc-
atively calling Cabrini-Green "the Soul Coast on the Gold Coast."
He also said he was acting as spokesperson for a new group named
the People's Organization of Cabrini-Green. The People's Organization pointed out that Cabrini-Green had as many inhabitants as most
suburban villages or towns, yet it lacked almost all of their services
and amenities. They demanded better street lighting. They wanted
childcare facilities, a medically staffed emergency room, and three
Olympic-size swimming pools—an average of one for every 5,700

people there. The organization also insisted that a black-owned company be hired to operate security at the housing project. "Black men are good enough to work as military police in Vietnam," Jackson proclaimed. "Why can't they be policemen right here in the Chicago police force?" Jackson sent a telegram to President Nixon, requesting federal disaster relief for Cabrini-Green. While the housing development had not "suffered either a natural or man-made catastrophe in the normal sense," Jackson wrote to Nixon, "the area must be considered a disaster area due to the potentially riotous conditions accompanying the resultant tension."

Nixon responded to Jackson's request, dispatching George Romney, his secretary of Housing and Urban Development, to Cabrini-Green. Romney, the former Michigan governor, had lost the Republican primary to Nixon two years earlier, going from presidential front-runner to also-ran in a few blundering weeks. A moderate Republican compared with Nixon, he'd reached on his own the same conclusion as the federal commissions that had studied the causes of recent inner-city unrest: segregation was destroying America. In his new cabinet position, he planned to use public housing to move African Americans out of urban ghettoes and into thriving suburbs. In Chicago, in 1969, a federal judge had ordered the CHA to do exactly that, ruling that the placement of the city's public housing developments in predominantly black neighborhoods fostered segregation and was in direct violation of the 1964 Civil Rights Act and the equal protection clause of the Fourteenth Amendment of the US Constitution. The class-action housing desegregation lawsuit against the CHA was named for Dorothy Gautreaux, who lived in all-black Altgeld Gardens, on the distant South Side. Although the judge ruled in favor of the *Gautreaux* plaintiffs, the case would be fought for years, all the way up to the Supreme Court. And Mayor Daley chose to put an end to almost all public housing construction rather than follow the court decree and build in white neighborhoods. From 1969 to 1980, Chicago would add little more than a hundred public housing units. Romney

believed he could withhold federal assistance to any suburbs unwill-
ing to add public housing as part of his Open Communities initiative.
Already an outsider in Nixon's White House, the housing secretary
decided to develop this desegregation program without revealing it
to the president. The plan would not go over well when it went public.

But first Romney toured Cabrini-Green. On a steamy August
day, two weeks after the cop killings, he and Jesse Jackson walked
the housing development. It was a strange scene. Romney moved at
a typical breakneck pace, while members of Operation Breadbasket and the People's Organization of Cabrini-Green formed a scrum
around the two men, pushing aside reporters and residents. Jackson
took the HUD secretary to the field where Severin and Rizzato had
been gunned down, pointing out the snipers' nests above. Romney
and Jackson hustled over to 1150–1160 N. Sedgwick, the Rock,
entering one of the apartments. With an entourage jogging to keep
up, they surveyed the expanse of concrete surrounding the towers, and
the tiny public swimming pool that locals dubbed "the Bathtub."
In Saint Dominic's Church, residents told Romney about the conditions in Cabrini-Green. The rats and the broken elevators, the single
mothers wiling away the days without hope, the gangs and the drugs,
the promises of jobs and services that never came, the police reminding them with a drawn gun or a curse to stay put on their island
of seventy acres. Romney was convinced. He announced that security
alone wouldn't be able to solve the problems at Cabrini-Green, and
he left. Jackson, finding himself alone in front of the reporters, declared the day an extraordinary success.

Yet conditions at Cabrini-Green only worsened. Out of punishment and fear, the city stopped regular trash pickup. Residents, too,
no longer thought about their home the same way. The murders and
the police response brought many to a breaking point. Distressed over
their own safety, elderly tenants were allowed to transfer to senior
housing elsewhere. More than seven hundred families moved out of
Cabrini-Green by year's end, a fifth of the 3,600 households, almost

all of them coming from the high-rises. Many people on the public housing waiting list opted to wait longer rather than accept a vacant unit in Cabrini-Green, which they'd heard so much about of late. Those who did move in were generally poorer and less stable than the tenants they replaced, and the portion of families on welfare and headed by a single parent both jumped to three-quarters of the total. The reality at Cabrini edged ever closer to the idea of it in people's minds. Mayor Daley visited the White House in May 1971, telling Nixon that racial fears had wrecked public housing in Chicago.

"In some portions there are 750 vacancies in Cabrini-Green. And white people won't move in," Daley said.

"What kind of thing is it?" Nixon asked of the public housing development.

"It's a white high-rise, though it's all black. But it's a high-rise with a high percentage of crime."

The urban planner Oscar Newman pointed to the design of high-rise public housing to illustrate his theory about the role that physical layout plays in abetting or preventing crime. Newman contended that superblock public housing lacked what he called "defensible space," and therefore represented the worst of postwar modernist architecture. The towers were without through streets or commercial establishments, creating dead zones with no sense of shared ownership or communal supervision. That's where crime was at its highest. In the words of Jane Jacobs—whose antimodernist vision from *The Death and Life of Great American Cities* guided a new breed of city planners—superblocks prohibited the self-regulating public safety from "eyes on the street" and the "normal presence of adults on lively sidewalks." In public housing projects, the majority of the violent crimes occurred in isolated and unmonitored stairwells, laundry rooms, and elevators. City officials joked that the most dangerous public transportation in Chicago was a Cabrini-Green elevator.

When HUD had discretionary funds available for improvements at its most troubled public housing developments, the sensationalized coverage of Cabrini-Green left little doubt where the money would go. "We wanted to find the worst of the worst public housing projects," a HUD official said. "The ones that because of physical deterioration, bad maintenance, problem families or overwhelming social conditions were really in trouble and need to be turned around. There were a number of factors considered. But in Chicago everyone knows that means Cabrini-Green." With the federal money that soon came, and in conjunction with funding from several state, county, and city agencies, the CHA began a $21 million security pilot program at Cabrini-Green to combat its structural deficiencies. The lobbies in four of the high-rises were reconstructed. Access to the buildings was curtailed, with guard stations built to block anyone from entering without first being screened. Bright lights were installed, as were vandal-proof mailboxes, polycarbonate glazed safety windows, and toilets in the lobbies, so children wouldn't relieve themselves in the stairwells. The four towers were outfitted as well with all-weather cameras that could capture images in the dark and with state-of-the-art closed-circuit TV systems for monitoring the elevators, stairwells, and perimeters of the buildings. The areas outside the high-rises were to be surrounded by chain-link and wrought iron fencing, brightly lit, and turned into grassy courtyards, so as to "increase the territoriality of individual buildings"—though these changes were never enacted. Neither was the promise to focus on "diversion as well as prevention," measures that were to include new social service programs for drug and alcohol abuse, for juvenile offenders, for school assistance, and for women to learn self-defense and problem solving strategies.

This investment in Cabrini-Green, as incomplete as it was, came at a time when other cities were already giving up on their public housing high-rises. Like Cabrini-Green, St. Louis's Pruitt-Igoe was a large public housing development, consisting of thirty-three identical eleven-story towers and a total of 2,870 units. It fell into incredible disrepair in a remarkably short time. The buildings were built poorly to start, without a sufficient number of playgrounds and with doorknobs and locks that broke upon first use; the St. Louis Housing Authority, too, was underfunded and inefficient, and Pruitt-Igoe residents rapidly became overwhelmingly poor, with the majority of their households headed by single mothers. And so in 1972, just eighteen years after the high-rises were completed, the Pruitt-Igoe towers were imploded with charges of dynamite. The televised image, with its mushrooming cloud of dust and debris, defined for many the popular notion of the public housing experiment—it had failed. The vision of decaying high-rises wasn't representative of the country as a whole, even if it reflected the reality in cities like Chicago and St. Louis: three-quarters of the nation's public housing units were not in high-rises.

No doubt there were many in Chicago who wished to do the same to the city's public housing towers. But the idea of demolishing Cabrini-Green seemed like a political as well as a practical impossibility. It would involve the displacement of some fifteen thousand poor, black Chicagoans. In the previous decades when slums were cleared, those displaced were sent into public housing. Now where could people from the projects be sent? After hearing so much about their criminal exploits in the news, what neighborhoods would be willing to absorb them? Sure enough, when suburban areas learned about George Romney's Open Communities proposal, they objected hostilely, and the program was scrapped before ever getting under way.

Instead of pursuing demolition, Chicago declared the security pilot program at Cabrini-Green a success. The first phase was completed at the start of 1977. Studies commissioned by the CHA showed that the renovated buildings experienced a 25 percent reduction in crime, a 38 percent drop in vandalism costs, and a 27 percent increase in tenant employment. The CHA reported that most of the empty units in the renovated high-rises were now occupied. The newspapers ran stories praising the effort: "Sunlight at Cabrini-Green"; "Cabrini-Green Revives." "The Cabrini-Green homes, despite an image of violence and dark despair, has become in the last two years a more comfortable place to live."

Hoping to build on the positive reports, the CHA marketed a new and improved Cabrini-Green, distributing an eight-page brochure of upbeat sepia photographs. The agency had in recent years sent mass mailings to families on the welfare rolls, trying to fill empty units in Cabrini-Green and Robert Taylor. But this pamphlet showed a nuclear family of four enjoying time in their refined living room, children frolicking in a playground and a pool, and rowhouses hedged by shrubs and blossoming foliage. "Cabrini-Green Homes is a community that most Chicagoans think they know all about, and what they think of it is usually all bad," the brochure offered as a corrective. "Located in the heart of Chicago's vibrant Near North Side, Cabrini-Green Homes is changing into a place where you can be safe and secure." The buildings were said to enjoy many of the same security features as the luxury doorman apartments along Lake Shore Drive, but with rents at a fraction of the cost and convenient off-street parking. "The Chicago Housing Authority expects great things from the new Cabrini-Green Homes, especially when families such as yours move there to help make that dream become a reality."

KELVIN CANNON

THE REALITY ON the ground in Cabrini-Green, especially for young guys like Kelvin Cannon, just entering their teens, seemed less than dreamy. The Blackstone Rangers, the South Side gang headed by Jeff Fort, were then undertaking a citywide recruitment drive that included the Near North Side. Students leaving Cooley or Waller High would find a group of Blackstone Rangers with bats waiting for them outside. "What d'you represent?" they'd ask menacingly. "What's your ride?" There were guys all over the streets near Cabrini wearing red berets and demanding to know what gang boys were in. The options were to say the Stones or to receive a beat down, though occasionally you'd get lucky—a teacher or a cop would appear, or the recruiter knew your older brother and would tell you under his breath to run.

On a summer day in 1976, when Kelvin was thirteen, he and his friends were playing strike out against their high-rise, the strike zone a spray-painted box on the sand-colored wall of 714 W. Division. Kelvin was in the outfield. A boy named Binky was pitching. Then five guys stepped between Binky and the batter. They were older, out of high school. By then the Blackstone Rangers had taken control of most of the Whites at Cabrini-Green. And now they were coming for Kelvin's home.

One of the guys pressed his shoulder against Binky's shoulder, their cheeks nearly touching, as the two of them turned together in the stomach-knotting slow dance preceding a fight. The older boy muttered incitements in Binky's ear—*You think you can beat me?* He pushed Binky hard, as if about to begin, but then engaged again, leading him in another series of turns. It was all ponderous precursor to the violence, as if the young man were searching for whatever would rouse him to action. It came when Binky tried to resume the game, stepping away to throw the rubber ball at the painted target. The Blackstone Ranger threw two quick blows to the pitcher's face. Then all was movement. The other four Stones with him piled on the punches and kicks. Kelvin and his friends rushed to help. They outnumbered the Stones, and they found themselves outboxing them as well. That's when one of the Rangers reached into his waistband, and everything went still. The pistol he pulled out was square framed and glinted in the light. He held it to Binky's face.

"I ought to blow your motherfucking head off." He raised the gun and cracked its butt against Binky's temple. The boys from 714 W. Division scattered.

When they reassembled later that night in the lobby of their building, they were still jumpy with adrenaline. Kelvin and his friends replayed every detail of the brawl, reenacting the moments that words alone couldn't capture. They'd held their own against twenty-year-olds. They weren't hooked up with any gang, but they weren't pushovers. For most of their lives they'd teamed up against other buildings for sports or snowball fights; three or four of them would head off together to the store to steal candy. "714 was the best in sports there was. So it wasn't a surprise to us that we came out on top in fighting, too," Kelvin would say. He was then lollipop framed, with a long, willowy body and a big round head of straightened hair. His dewy eyes and wide, fixed grin didn't project menace, but there was little that he feared. He would gladly defend his building. "Bear arms," was how the group of them put it. Some of their older brothers were back from Vietnam, and the boys had access to guns. Really they only wanted to show that they had guns just like the Stones had guns. They weren't going out there like them, terrorizing people. They weren't planning to shoot. But over the next months they did have gunfights with the Stones, and they took them on fist to fist, too. They managed to push the Stones back to Sedgwick Street, near the Marshall Field Garden Apartments. The Blackstone Rangers were forced out of most of Cabrini-Green. Kelvin liked to say that their resistance changed the course of history at the housing project forever.

While they'd overcome the Rangers, the guys from 714 W. Division started skirmishing with the boys from the surrounding high-rises. The groups of friends from those buildings had banded together in the same way. They bore arms to keep other teens from storming their entranceways. Guys from 660 W. Division or 1230 N. Larrabee might come over to 714, mess with their girls, disrespect them. A punch would be thrown, and it demanded a response. If they pushed up on you, you went to their building to body slam one of them. One day, a bunch of boys from 630 W. Evergreen snuck into Kelvin's building and trapped him on the ramp. They came at Kelvin from the stairwells on either side of him. Kelvin figured it better to deal with half of the boys than let the entire group whale on him, so he ran toward one of the stairwells. Several boys grabbed him and tried to pull him down the stairs. Kelvin swung wildly. One of the guys shouted, "I got him!" But Kelvin landed a blow to his face. Then Kelvin's friends from his building heard the hollering and came to help, and the intruders fled.

Those boys from 630 W. Evergreen called themselves the Insanes. In another surrounding high-rise, they were the Imperial Pimps, and in two of the Reds there were the Outlaws and Renegades. Even 1230 N. Burling, mostly neutral when it came to fighting, had the Black Pearls. Kelvin and his buddies in 714 W. Division decided they were the CCs, for Chocolate City. To them that sounded cool. But they started to think the name too soft. "We wanted to sound as hard as everyone else," Kelvin said. They changed from Chocolate City to Imperial Gangsters.

The fall of that same year, not long after the start of school, the leader of the Gangster Disciples at Cabrini-Green summoned all these different peewee gangsters from the white high-rises. They were told to meet after class one day, at three thirty, behind their middle school, in the park named for President Lincoln's secretary of war, Edwin Stanton. Sixty of them gathered there, the boys from each building huddled in their different cliques. Then Bo John arrived, flanked by two of his lieutenants. His real name was Cedric Maltbia, and he was twenty and recently out of the penitentiary. He wore a sleeveless T-shirt that showed off just how much he'd bulked up in prison, and from his neck hung a large, gold six-pointed star—the emblem of the Disciples. Every kid there knew about Bo John. He lived in the Whites, too, and some of the boys had played on Big Wheels with his younger brothers. One of Kelvin's older brothers ran with the Disciples; he'd recently been violated by the gang, beaten with bats, for shooting heroin. Kelvin accepted that his brother deserved the sanctioned punishment, since he'd broken the gang's bylaws about drug use. Kelvin was willing to hear Bo John out.

"I call it for the Folks over here," Bo John began by way of introduction. He praised the guys from 714 W. Division for fending off the Blackstone Rangers. He said it took courage to take on grown-ups. Kelvin and his friends blushed with the thrill of the important man's attention. But then the Disciples leader scolded the lot of them for tearing up the community. "Y'all acting just like the Stones, battling one another." He said they'd all grown up in the Greens. He wanted them to imagine how strong they'd be if they didn't waste their energies beating on one another. Other gangs were corrupt, he told them. But the Disciples fought for what was right. They lived by a code, by rules. They didn't jump on people for no reason, and they battled heads up. Everyone in front of him had already proven himself worthy of being a GD. "You all might as well ride together," Bo John said. "I'd like y'all to rise up under me."

There was nothing Kelvin loved so much as a gangster movie. After he saw *Once Upon a Time in America* a few years later, he and his crew started calling each other by the characters' names—Noodles, Max, Patsy. Kelvin became Big Frank, though mostly he was just Cannon, which people assumed was not his real name but a nickname he'd earned through some display of prodigious violence. One of his all-time favorite movies was *The Warriors*. When he first watched it at the McVickers Theater, in the Loop, he realized Bo John was just like the gang leader at the start of the movie, the one who preaches unity to all of New York's street gangs, right before he gets shot, telling them that the future is theirs, if they can count—they had strength in numbers. They needed only to overcome their divisions, to join together, talking about "CAN YOU DIG IT!" That's how Bo John had talked to them behind Schiller school, that the Gangster Disciples could take over all of Cabrini-Green—the rowhouses, the Whites, the Reds along Larrabee, known as the Boulevard, and those on the southern edge of the development as well, the Wild End. All they had to do was follow him.

It was a message most of the middle schoolers were eager to hear. By then, Jesse White had been tapped by George Dunne to run for the Illinois General Assembly. The campaign posters showed White in his Cubs and paratrooper uniforms, strong jawed and stately. Cabrini-Green offered a solid block of Democratic machine votes, and White won the election. He continued to teach at Schiller and attend every Tumblers performance, hundreds of them each year. But with regular trips to Springfield he had to cut back on his other extracurriculars at Cabrini-Green. Kelvin thought sometimes that White's start in the legislature marked the turning point for Cabrini-Green. "Mr. White was a blessing from God to us here," Kelvin would say. "But he had to go on to politics and help more people. That's when everything went bad. We didn't have nothing else to do. Maybe more of us would have stayed in school if Mr. White had remained more involved. Maybe a lot of people wouldn't have gone to jail or been killed." For Kelvin and many others there, it felt like they'd lost another father.

Reginald Blackmon, one of Kelvin's best friends, would later say that Bo John was the first man who ever showed him love. Kelvin saw Bo John more as a big brother. His own father had left for the South Side; his brothers were doing their own thing; he was the oldest male in his apartment, with his mother and younger siblings. Kelvin felt it was time for him to grow up. "At Cabrini-Green, you didn't have to wait until you were eighteen to be a man," he'd say. "You could be a man at twelve or thirteen." Bo John gave them money each week; that was better than turning flips with Mr. White. And truth be told, by the time Cannon and his friends were thirteen, fourteen, it was no longer too cool to hang with Mr. White. They didn't want to be seen wearing the white suspenders and red tams of the tumbling team. Of the hundreds of Boy Scouts under Jesse White, the older brother of Reginald and William, Richard, was the only one any of them knew who tried for Eagle. Like Dolores Wilson's oldest son, Che Che, Richard ventured out of the neighborhood to Lane Tech, the prestigious magnet high school a bus ride north. On the mornings when his mother couldn't muster the bus fare, Richard was assisted by the drunks along Halsted, who pooled their nickels to get him to school. He stayed clear of the gangs and went to college, in southern Illinois, later becoming a lawyer and then an educator at a public high school on the South Side. "I adopted Jesse White's belief," Richard would say. "I don't listen to what's not possible."

Richard said that his little brothers and Kelvin were wonderful, bright-eyed boys, athletes, math whizzes, goofs around girls. Maybe they buzzed on reefer or cheap wine, but they didn't know anything about cocaine or heroin. These boys from 714 W. Division were still in middle school. Then all of a sudden Richard didn't recognize them. They'd transformed. Just a few years earlier they believed in witches haunting the Ogden Avenue Bridge. Adults didn't even bother with that tall tale anymore. By comparison, a spectral gypsy no longer seemed like much of a threat. Cannon and the other young gang-bangers now imagined that they were the ones to be feared at Cabrini. And Cannon believed that it was through such fear that he, too, could provide a little bit of order to the chaos of life there.

5

The Mayor's Pied-à-Terre

DOLORES WILSON

IN HER JOB with the Department of Water Management, Dolores Wilson helped process the paychecks for the employees of two pumping stations; she filed, typed, and sorted the mail; she answered the phones for three purification labs and the head of microscopy. She was considered a good worker, at one point the department's employee of the year. Yet even with all her responsibilities, Dolores often found herself with little to do. This was, after all, a patronage job under Chicago's Democratic machine. One afternoon, she knocked on her boss's office door. Would it be okay if she read during the downtimes? It was fine as long as she didn't look too conspicuous; she shouldn't prop her feet up on the desk or anything like that. So Dolores demurely plowed through romances and inexpensive paperbacks. She devoured the Mandingo series, about slavery and sex on an Alabama plantation. She finished one long Russian novel after another.

Dolores tended to worry herself sick over the suffering of others, even animals. "It's like a phobia," she would confide. If she saw a lion catch a zebra on television, the zebra's cries would stay with her for days. At Holy Family, her church at Cabrini-Green, she'd pray out loud for people in Africa and Siberia. She'd pray for the owners of pit bulls to have a clean heart, a kind heart, and not to enter their pets in fights. She'd go weeks without reading the newspaper because she couldn't handle stories of people in Vietnam, Cambodia, or wherever else they were being bombed and gassed and massacred. Hubert told the children not to share bad news with their mother, and sometimes she'd learn about a death months after it occurred. She was haunted especially by a line she'd read in Dostoyevsky, about a society being judged by its prisons. Kenny, her youngest, was once sent to the Audy Home, the city's juvenile detention center. It was winter, and the guards made the boys strip down and shower, lining them up wet and naked by the open windows for hours as their bodies convulsed from the cold. She couldn't fathom why one human being would do that to another one. Lately, she'd been imagining the prisoners who were trapped in solitary confinement. She realized that at any moment, in some godforsaken penitentiary, there were people shut inside cramped boxes, without windows or contact with others. She learned that inmates might be placed in the "hole" for a week at a time. Were they allowed visitors? Did they receive letters? Were they denied every kindness?

Along with the organist at Holy Family, Gloria Johnson, Dolores decided to start a pen pal program for the incarcerated. Jesse Jackson had moved on from Operation Breadbasket, founding his own Chicago-based organization, Operation PUSH, for People United to Save Humanity. From the PUSH folks, Dolores received a list of twenty-five inmates, and she wrote each of them, explaining who she was and what she was doing, promising that she wasn't the warden trying to trick them into revealing information. She then reached out to fellow church members and neighbors, finding a letter writer for every prisoner. Gloria took a guy named Jerome, who was serving a long sentence. Dolores's daughter Debbie wrote another prisoner. For herself, Dolores picked a man named Maurice Slaughter. "With a name like that, I figured he could use a friend," she said. Those first twenty-five incarcerated men told others about her. Soon Dolores was receiving letters from prisons all over Illinois and then throughout the country, with the number of pen pals reaching into the hundreds.

Once, a man from Stateville, the penitentiary less than an hour out-side of Chicago, called collect to ask Dolores if she would stay on the phone and listen to one song with him on the radio.

Before long, she was jumping into Gloria's rickety Volkswagen first thing on Saturdays. In Chicago, being sent "downstate" meant going to prison, and the two of them would drive to the different Illi-nois towns that housed the penitentiaries: Dixon, Joliet, Menard, Pon-tiac. In wintertime, they'd wear their coats and boots the entire ride since the heat in the Volkswagen didn't work. During visiting hours, they'd talk with the inmates, and when possible they brought with them the prisoners' families. They had cards printed up calling them-selves "Prison Problem Consultants." They'd show the cards to the guards with formality, and because they were "consultants," they no longer were limited to meeting with just one inmate on each trip. The more the men told Dolores about their experiences in prison, the more distressed she became. Guards hit them with chains and handcuffed them to their cells in standing positions; the inmates were forced to sleep on concrete floors. Many of the black inmates had turned to Is-lam in prison, and they weren't allowed to worship. Once, she arrived when the prison was on lockdown: the men rattled their metal cups against the bars as the guards blasted them with water from fire hoses.

Dolores now thought about the prisons nonstop, images of depri-vation and torture coming to her as she was falling asleep or riding the bus to her job. She cursed the guards and the wardens, the judges and the lawyers, the police and whoever else determined the fate of these men. She started writing letters to the Chicago newspapers, composing them on her lunch breaks. Writing about the Menard 38, inmates who barricaded themselves in the commissary in 1973, holding a guard hostage, she explained that the men were trying to bring attention to the inhumane conditions at their penitentiary. She argued that wardens and guards were effectively sentencing prison-ers a second time by condemning them to physical pain. With no effort to rehabilitate them, the incarcerated were stripped of their dignity; they were being taught to hate and distrust, dooming them before any release back into society. "We, the so-called intelligent and educated, are allowing our men to be made into animals, because they're being treated as such," she wrote in a letter published in the *Defender*. "We are the hopeless, because none are so blind as those who refuse to see and act against injustice!!!"

She published so many letters in the paper that the *Defender* gave her a byline, calling her a "prison correspondent." The newspaper described her as a full-time stenographer, a wife, a mother, and a grandmother who was using her nonexistent "spare" time to communicate with those in Illinois prisons. The accompanying photograph showed her at forty-four, with cornrowed hair, heavily arched brows, and large hoop earrings. It's her eyes, though, that stand out the most; they look past the photographer, the brown in the irises dancing with a private joke and something like judgment. One of the prisoners got ahold of the article and painted a portrait for Dolores based on the photograph, presenting it to her the next time she visited.

As overprotective as Hubert was, he didn't feel jealous of his wife spending her precious time on these lonely, incarcerated men. Gloria even ended up marrying her pen pal, saying she'd wait the twenty years for Jerome to get out. Hubert was charitable in his own right. Dolores came home once to find him feeding a homeless man at their kitchen table. What Hubert didn't like about the prison work was that the men were criminals. They'd made their beds, he'd tell Dolores, let them lie in it. They didn't deserve her empathy. He and Dolores would bicker about it, but the arguments never lasted. They'd been married for more than twenty-five years, the majority of their lives. They threw a silver anniversary party at Saint Joseph's, inviting all their friends and relatives. They went boating together, horseback riding. And each year, Dolores and Hubert coordinated their vacation days so they could go to Jamaica during the cheaper off-season; thirteen hurricane seasons in a row they made the trip, always booking a room in the same Holiday Inn.

Hubert was promoted to an assistant head janitor position at Cabrini-Green, and Dolores, upon hearing the news, cried. The new job was in a different section of the housing complex, the Whites, and they would have to move across Division Street into 1230 N. Burling. Dolores didn't want to go. She had everything in their 1117 N. Cleveland apartment exactly how she liked it. All the interior decorating was in place. She knew every neighbor. "It isn't where you live, it's how you live," Dolores would say. "I could move into one of these high-rises overlooking the lake and everything. Until a part of me is in there, it's just an apartment. A home isn't a home until you've been in it awhile."

On the morning of the move, as a passive protest, Dolores refused to help, leaving early for the water department. That was fine with Hubert. He carried their belongings north of Division Street and up to the fourteenth floor of the Burling high-rise. He unpacked, putting each item in its proper place. He hung his wife's yellow curtains. He outfitted the kitchen with an upright freezer, an electric dishwasher, and a washer and dryer, all the appliances in the same color of harvest gold. In the bathroom he painted the walls peach with a star pattern, so it almost looked like wallpaper, and installed a shower door with a swan design. The new apartment was bigger: four bedrooms instead of three, and it had a bathroom and a half. When Dolores returned from work to the unfamiliar surroundings, she couldn't help but like what she saw. Neighbors she didn't know were inside the apartment, too, admiring the setup. Dolores asked Hubert what all these people were doing there. He said they were envious: "Because we're more fortunate than they are."

"How am I more fortunate?" Dolores asked, unable to stop smiling.

IN 1976, RICHARD J. Daley was visiting his doctor for a scheduled appointment. That morning he'd celebrated the opening of a new gymnasium in a white ethnic neighborhood in the city's southeast corner, declaring, "This building is dedicated to the people of this great community. They're making Chicago a better city, because when you have a good neighborhood, you have a good city, and this is a good neighborhood." In the doctor's examination room, Daley had a heart attack and died. He was seventy-four and had controlled the city as mayor for twenty-one years. During a week of closed-door meetings, the party leaders debated who would fill the vacancy, finally picking a Democratic machine loyalist named Michael Bilandic, a mild-mannered alderman and corporate lawyer from Daley's Bridgeport neighborhood. "Mayor Bland," as he was sometimes called, struggled to quell labor disputes with butchers, gravediggers, and musicians with the Lyric Opera. In the winter of 1979, just weeks ahead of the election for his first full term, a blizzard struck the city, nineteen inches of snow, and the "city that works" couldn't even clear the streets. Bilandic was edged out by a challenger he'd fired from a job as head of the city's consumer affairs division. Jane Byrne was Chicago's first—and, as of 2018, still the only—female mayor.

Byrne ran as an anti-machine reformer, but she'd also served dutifully under Old Man Daley. The presidential campaign of John F. Kennedy, in 1960, inspired her to get involved in local politics. Recently widowed with a baby daughter, Byrne was granted an audience with Daley in his fifth-floor office, and he instructed her to speak to her alderman. The code of the Chicago machine is often summed up by a koan uttered to a future US congressman and federal judge named Abner Mikva when he was twenty-two and trying to volunteer at his South Side ward office; the boss there removed the nub of his cigar long enough to spit out, "We don't want nobody nobody sent." Although Byrne didn't even know her alderman's name, the Man on Five himself had sent her; she was given a job ringing doorbells. Four years later, Daley set her up in a government post, with the Chicago Committee on Urban Opportunity, the local agency overseeing Lyndon Johnson's War on Poverty. The money from Washington passed through her office before making its way to Cabrini-Green and other impoverished neighborhoods, so that every paycheck for every job funded by the Feds had the mayor's name on it. Chicagoans needed to know that they owed their livelihood not to Johnson or his Great Society but to Daley. In Byrne's memoir, *My Chicago*, she recalls Daley telling her, "When I appoint personally, that person is hand-picked by me—and IS mine. You owe your loyalty directly to me; you answer only to me." He later put her in charge of the Department of Consumer Sales, Weights, and Measures, Daley declaring that she would be the first woman to lead a big-city agency.

With Byrne as mayor, Cabrini-Green again dominated the news. The management-consulting firm Arthur Young and Company, hired to assess the lasting benefits of the millions spent on security there after the Severin and Rizzato killings, found that there were none. The firm cited unemployment as the most pressing problem, but noted that the CHA had pulled back on routine maintenance. When security cameras broke, they weren't replaced. The number of main-tenance workers at Cabrini-Green had been cut from nineteen, in 1977, down to just six. Hundreds of apartments were badly in need of repairs, but the agency, facing rising deficits, did little as condi-tions worsened. The vertical patrols started in 1971 that were sup-posed to send officers through every floor of each high-rise, securing the buildings from top to bottom, had devolved into do-nothing units. The officers stayed in their cars and never entered the high-rises. One officer admitted that they spent the nights mostly playing cards and watching TV: "Anything but serve and protect." When Byrne officials showed up at Cabrini unannounced, they discovered that almost half the patrolmen assigned to the detail hadn't bothered to come in for their shifts.

The overcrowding and growing unrest in the state prison system, as well as the attention it received from Dolores Wilson and other citizen and professional "prison problem consultants," led Illinois governor James Thompson to initiate an early-release parole pro-gram. As a victory, it was a Pyrrhic one. Since Cabrini-Green and other beleaguered public housing developments in Chicago had vacancies, that's where large numbers of ex-offenders were sent. These were people who needed shelter and a fresh start, but also required extensive guidance and counseling; parolees were supposed to be monitored and mentored. Congress had recently passed legislation giving those involuntarily displaced or in "substandard" living conditions greater access to public housing, and making it much harder for working-class families to secure a unit. The consequence of these well-intentioned policies, however, was that society's dire problems were being dumped into neglected high-rise projects—it was less a solution than the making of a crisis. By 1981 there were seventy-seven former inmates with known gang affiliations paroled to Cabrini-Green from state prisons, another three hundred from the Cook County jail, and an impossible-to-determine number living there as squatters. Oftentimes criminals never even kept a Cabrini apartment, instead treating Cabrini-Green as a police-sanctioned vice district, a place to do their dirt before returning to wherever else they called home.

The number of parolees contributed to a crime wave at Cabrini. There were thirty homicides there from 1978 to 1981. In the first three months of 1981 alone, eleven people were murdered and thirty-seven others shot. That January, Cedric Maltbia, the Gangster Disciples chief known as Bo John who'd recruited Kelvin Cannon, was killed in a Cabrini high-rise. Two women, sisters who'd moved to Chicago from New York, had come to the building to shoot a man who'd beaten and robbed one of them. Bo John happened to be in the guy's apartment, so the sisters shot him, too. With the Disciples leaderless, the fight for succession added to the mayhem. One of the ensuing killings occurred in the ground-floor rec room of 1230 N. Burling, Dolores Wilson's new home. Thirty people had gathered there on a Monday night in March to hear a musical group, young men who called themselves the Electric Force Band. Junior Miller, who played bass in his father's church, was on guitar. Jimmy and Ronnie Williams, who lived in a neighboring high-rise, played drums and keyboards. Demetrius Cantrell, whom everyone called Sugar Ray Dinke, for his boxing skills and diminutive size, had taught himself guitar by watching the old bluesmen on nearby Mohawk Street. The lead singer was a twenty-one-year-old named Larry Potts, who could sound like Frankie Lymon or Little Anthony. They covered radio songs, mostly funk and soul—Cameo, Rick James, Con Funk Shun. They played Jerry Butler and Curtis Mayfield, of course, giving the Cabrini legends their due. Because Hubert Wilson knew the young men from the Corsairs, his drum and bugle corps, he'd help them set up on the ramp or in an empty apartment. They might charge a quarter or fifty cents for people to come and listen. If it was hot out, they plugged in an orange extension cord, connected it to another one, and then another one, and played outside the building, turning the blacktop into a dance floor.

On that March night, Larry was singing when someone passing by outside stuck a .357 Magnum through an open window in the rec room and fired several times. A boy in the audience pushed Hubert Wilson to the ground and lay on top of him. One bullet hit a six-year-old in the right thigh, and another shot ricocheted off a wall and struck a fourteen-year-old girl in the leg. Both children would be okay. Larry Potts was shot through the back, and died that night at a North Side hospital. The shooter, a twenty-four-year-old named Jerry Lusby, a Cobra Stone from 1150 N. Sedgwick, on the other side of Division Street, told police that he mistook the singer for a rival gang member.

The next morning as Mayor Byrne was getting ready for work, drinking coffee and putting on makeup, she heard on the radio the news of the singer's death. Not his name or who he was. The reporter simply broadcast that there'd been another killing at Cabrini-Green. Byrne lived on the forty-third floor of a luxury high-rise less than a mile east of Cabrini. She could look out her window on the Gold Coast and see the housing project's palisade of red and white towers.

When Byrne's grandfather emigrated from County Mayo, Ireland, to Chicago, in 1888, the first place he lived was in Little Hell, the Irish ghetto that was the future site of Cabrini-Green. Her grandfather's older brother, who'd made the journey four years earlier, warned him to stay clear of the gangs that terrorized the neighborhood. A century later, gangs there were still causing havoc. Byrne decided that as mayor she was obligated to do something about the violence. Eleven days later, she announced that she would take up residence at Cabrini-Green. "I will keep that Cabrini-Green apartment in the way that many suburbanites keep a downtown or in-town apartment," she said. "I will consider it as a place to go on some nights and not on others. I will not give up the apartment so long as I am mayor."

She picked as her new home what she'd heard was the most troubled of all the high-rises, the building where Larry Potts's killer lived, 1150–1160 N. Sedgwick, the Rock. When the building's manager started showing apartments to Byrne's advance team, the first "vacant" unit he opened had a family living in it. He used his keys on a second officially empty apartment. Unofficial occupants lived there as well. By then the overall population at Cabrini-Green had fallen from a high of somewhere around 20,000, in the 1960s, to 14,000. But the housing authority estimated that another 6,000 people likely stayed there who weren't on a lease. The head of the CHA guessed that 300,000 lived in the city's public housing, more than double the official count and a tenth of Chicago's total population.

A reporter who visited the high-rise before the arrival of the mayor's cleaning crew described the fresh urine pooled in one of the two elevators, the graffiti covering the walls—"Disciples Kills all Stones"—and the fear he felt in the narrow corridors and stairwells. The toilets were out of order in the lobby. The screen door on the mayor's new fourth-floor apartment was punched in and lolling off its hinges. Second graders at Cabrini-Green's Byrd Elementary—a school created to relieve overcrowding at Jenner—sent letters to Byrne urging her to stay away: "A lot of black people live here and you are a white person"; "Roaches and rats might drive you crazy"; "You may be shot, stab or assassinated." In 1978, Byrne had married Jay McMullen, a city hall reporter with the tippling insouciance of an emcee at a Dean Martin roast. When McMullen first visited the Rock, the elevator taking him up to his apartment broke down, stranding him and eleven others between floors. When freed, he walked the four flights, the light bulbs missing in the stairwells, trash piled up around the garbage chutes at each landing. "It ain't the Ritz," he told the reporters with him, but he wasn't too worried. "I've slept in some pretty unusual places in my life."

The couple moved in on March 31, 1981, arriving in a limousine and hurried inside by a security detail of sixteen. The apartment next to Byrne's was cleared so that two guards were stationed nearby at all times. Bulletproof glass panels were installed on her windows. President Reagan had been shot the day before, in an attempted assassination. Byrne had received death threats as well. "While I was mostly philosophical about such threats," Byrne wrote in a diary of her stay that was published in the *Sun-Times*, "Jay was grimly humorous, stating, 'You'll be safer in Cabrini. The place has such a bad reputation most assassins will be afraid to go there.'" She added, "And now it is our pleasure to take turns using a hand shower that our guards attached to the bathtub faucet. They don't live very fancy here at Cabrini."

Byrne was engaged in a brazen political stunt, a ploy to jumpstart declining poll numbers midway through her first term. She'd won the black vote over Bilandic, but once in office she'd reduced the number of African Americans in key civic positions. She'd installed a black interim police chief but given the permanent job to Richard Brzeczek, a white veteran of the force. She removed two of the five African Americans on the eleven-member school board, replacing them with whites who were openly opposed to desegregation through bussing. Byrne had pivoted, hoping to shore up support among whites, but the backlash was fiercer than she expected. In

Chicago's zero-sum racial politics, each of these moves was treated as a declaration of war. Relocating to one of the country's most infamous black ghettoes would demonstrate that she cared about issues affecting the African American community.

As political theater, it was compelling stuff. Byrne was a small woman, heavily rouged, with a blond bob and a mouth that seemed perpetually pursed with nervousness. She was partial to mink coats, ruffles, and pastels. The local and national news ran with the story of this tough Irish woman living in the notorious housing project. She was like Kurt Russell's character in that year's *Escape from New York*, entering the ruins of the postapocalyptic city, except she wore a purple suede jacket and a pink skirt. "Mrs. Byrne is crossing the invisible but powerful line which has always separated the haves from the have-nots," wrote the *Defender*. "It is a stunt whose redeeming political symbolism elevates it to the lofty heights of civic and moral responsibility."

To her credit, Byrne understood the practical effect the media attention would have on Cabrini-Green. She occupied the seat of power, and she could bring that power with her to public housing. "Wherever a mayor goes, there seem to be city services galore," she said. Once there, she launched a sports initiative for the thousands of local youth, beginning construction on three new baseball diamonds and a football field on the site of the shuttered Cooley High School. Two other parks received upgrades. Potholes were filled and cars towed. Workers from the city's streets and sanitation department swept gutters and picked up trash. Sod was planted (but it all died). Sewers were repaired. Plumbing and heating systems in the buildings received upgrades. Byrne created a new food cooperative at Cabrini. She sent staffers from Human Services into the high-rises to counsel truants, work with victims of crimes, and help alleviate domestic conflict. Seven area liquor stores were closed, the city citing them for a range of electrical or structural violations, and hundreds of families were evicted, some for illegally housing parolees suspected of gang activity. Byrne opened a new misdemeanor court at a nearby police station to hear what was expected to be a torrent of new cases. A retired army major general, who'd commanded the Green Berets, was given the position of directing all security at Cabrini. A special task force of fifty federal agents from the Bureau of Alcohol, Tobacco, and Firearms was assigned to stop the trafficking of guns there. For the 150 Chicago cops who participated in a Sunday raid of the purportedly vacant apartments at Cabrini-Green, she gave each of them white envelopes with $50 inside, money from her political fund, instructing them to treat their wives or girlfriends to a nice meal; Byrne also awarded $800 to six officers who solved a Cabrini murder. Police crowded the streets; firemen and paramedics entered the towers without fear. More work went on at the housing project in two weeks than had occurred in the previous two years.

Byrne's Cabrini sojourn led every TV newscast; slightly different installments appeared in the morning- and late-edition newspapers. The *Sun-Times* ran daily opinion polls: "Will Mayor Byrne's Cabrini-Green move make a difference?" "If you were Mayor Byrne, would you move into Cabrini-Green?" The city council passed a resolution praising her decision to take an apartment there, and other city politicians announced their own intentions to stay in public housing. When New York City mayor Ed Koch was asked if he, too, would consider a similar move, he said he'd grown up poor and had no desire to go back.

"Rumors, roaches, rats and gangs are the curse of Cabrini," Byrne wrote in her *Sun-Times* diary. She mused on the subject of the cockroach, how she'd developed the reflex to sweep walls and let faucets run before using them, so much so that when she left town on business, spending a night at a luxury hotel in Manhattan, she found herself doing the "cockroach check." "Did you remember to pack the Raid, dear?" her husband asked drolly. She recounted going to bed to the *CBS Evening News*, just as she and Jay did most nights, though now they were in a Cabrini-Green bedroom and Walter Cronkite was talking about *them*. In one diary entry, Byrne described standing at her apartment window as white joggers waved and blew her kisses from Division Street: "It was like seeing the first robin in the spring. We hope more will come by and not be afraid." A poll found that 60 percent of Chicagoans would vote for her if the election were held then, with 27 percent saying her Cabrini stay changed their vote. Two-thirds of those interviewed felt she was trying to solve problems at Cabrini-Green, and nearly three-quarters thought the move sincere.

In many struggling sections of Chicago that were not Cabrini-Green, community leaders couldn't comprehend why the city was funneling its scant resources into a single seventy-acre plot of land. There were dozens of other public housing developments in the city, and numerous neighborhoods in need of services and revitalization, most of them without the advantage of being a few blocks from the mayor's Gold Coast home. Many Cabrini-Green residents also resented Byrne's presence. They accused her of creating a police state; residents reported being stopped and frisked five times whenever they walked from a building to the corner grocery. Hundreds were arrested, almost all for misdemeanors, with just about every case eventually dismissed. Elax Taylor, who'd operated the 911 Teen Club in the basement of his high-rise for decades, received an eviction notice because his seventeen-year-old son was found with marijuana. Even Police Superintendent Brzeczek admitted that his officers were treading a "very, very fine line between maintaining order and becoming oppressive."

The activist Marion Stamps was especially vocal in her criticisms of the mayor. Her community center, Tranquility-Marksman, sat across Division Street from Byrne's new apartment, and she had a close relationship with the families in 1150–1160 N. Sedgwick, the young men even bestowing on her the honorific "Mama Stone," for the support she'd given youngsters who later became Cobra Stones. Stamps was a short thirty-five-year-old with a round, expressive face, often framed in oversize oblong glasses. But with her bullhorn voice and intensity, she gave the impression of someone of much larger stature. She was born in Jackson, Mississippi, in 1945, and as a girl she picketed the segregated public library that wouldn't lend her books. Medgar Evers lived close by, and the civil rights leader trained her as an organizer. When she was seventeen, she moved to Chicago, and in 1965 she landed an apartment in the 1230 N. Burling building. Public housing, she said, was "a godsend, a blessing, compared to the slum housing I was living in before. For me the move to Cabrini-Green represented something bigger and better." It was also a place, she would say, where "social workers question your man- and womanhood and think nothing of it. Politicians make promises of jobs and welfare checks for your votes, or no jobs and no welfare check if you don't vote the way the precinct captain has dictated. Then we have Chicago's finest, the police who only serve and protect property and property rights." She blamed the "street organizations" as well, "the misguided black folks among us who sell dope to our children, who intimidate and force our children into gangs."

All of this compelled her into activism. At Cabrini, she raised five daughters and headed up the Chicago chapter of the Housing Tenants Organization. She ran a program for expectant mothers, since the infant mortality rates there were on par with those in the third world. She pressed for a new school to be built in the community, one she helped get named for the slave-turned-abolitionist Sojourner Truth. And she despised the idea of Mayor Byrne as a "white savior" coming to Cabrini-Green to rescue the poor black folk. With a flair for the incendiary, Stamps compared Byrne to the Ku Klux Klan and told news outlets that life at Cabrini-Green with the mayor there was like living in a concentration camp or a South African township under apartheid. "When you are not free enough to speak up or go out of your house, you are already victim to a form of death," she said.

The tensions came to a head at the mayor's "Spiritual Easter Celebration," a daylong event held in the third week of Byrne's stay at Cabrini. A giant white cross was erected on Division Street, and a choir sang along with Byrne that "God's got Cabrini-Green in His hands." The event featured a Ferris wheel, men playing the bongos, free cotton candy, and circus rides. Reggie Theus of the Bulls, Chet Lemon and Minnie Minoso of the White Sox, and members of the Bears and the Chicago Hustle, the Women's Professional Basketball League franchise, all spoke from a stage about hope and revival. Byrne was introduced by one of her officials as "the newest and one of the truest residents of Cabrini-Green." In her memoir, Byrne said she responded to the event's low turnout by confronting the gangs directly, accusing them of scaring children away from the festivities. Division had become an actual dividing line between the Cobra Stones in the Reds on the south side of the street and the Disciples in the Whites on the north side. Even Disciples from the Whites and the Reds saw each other as rivals. Pleading from the stage, Byrne implored children to go back to their high-rises to get their friends, since they'd surely hate to miss the free food and games.

As the mayor sang an off-key "Easter Parade," protesters waved placards, chanting, "We Need Jobs, Not Eggs." Byrne contended that the demonstrators were all gang plants. But they included a range of Cabrini-Green tenants and other activists, among them Marion Stamps. Her organization counseled dozens of residents who were evicted when Byrne moved into Cabrini-Green, and they managed to reverse almost every single case. Stamps stood alongside members of Slim Coleman's Heart of Uptown Coalition, a group that blamed the mayor's eviction policies for creating tensions between poor blacks and poor whites, many of whom were former Appalachian coal miners who'd moved to Chicago's North Side in pursuit of a better life free of black lung disease. Stamps said that the placing of the white cross in their black neighborhood felt like psychological warfare. She pointed out that Byrne managed to employ every convention of the white colonizer subduing a native population, appeasing them with religion, sports, petty entertainments, junk food, and trinkets. At some point the police decided they'd heard enough. As officers cuffed and hauled the protestors away, a lawyer tried to intervene. "What is he being charged with?" he shouted. "Who's the arresting officer?" He was thrown into a paddy wagon as well. All the while, one Cabrini resident shrieked, "Assassins!"

Byrne left on a California vacation the following week, deciding then to end her residency at Cabrini-Green. In the twenty-five days that she lived in an apartment there, only one person had been shot. The crime wave had subsided. "We never will leave Cabrini and neither should anybody else," Byrne wrote in her final diary entry. The mayor did return to Cabrini-Green occasionally. She showed up later that year when two Cabrini teens opened the area's only news-stand. Over the summer she led the ribbon cutting of a new athletic complex named for Severin and Rizzato, the plaque stating, "This field is dedicated to all people who wish to live together in brotherly love." A hundred dignitaries were in attendance, including family members of the slain officers. Byrne came out for a couple of baseball games, and her husband coached a team there for a few seasons. Mc-Mullen bought children mitts, and when the grass went weeks without a cutting, he called the park superintendent to get it done.

But Byrne's bump in popularity from the Cabrini move was short-lived. Within a day of her leaving, news outlets were already going with the story of the housing development's intractable problems: "Residents say services left with her." The sixteen-member security detail set up to guard the lobbies and monitor closed-circuit televi-sions was disbanded; the CHA said funding dried up.

Byrne also undermined the goodwill she'd garnered with African Americans. For starters, she refused to get rid of Charles Swibel, who'd been abusing his position as head of the CHA since 1963. Swibel was one of Byrne's biggest fund-raisers and closest advisers, and the two of them often rode together in her limo. From 1978 to 1982, nine different reports by auditors and consultants found that the CHA was in shambles. The ninth of these reports was conducted by Oscar New-man, the author of *Defensible Space*. Yet Newman concluded that the problems in Chicago went way beyond bad architecture: "In every area we examined, from finance to maintenance, from administration to outside contracting, from staffing to project management, from purchasing to accounting, the CHA was found to be operating in a state of profound confusion and disarray. No one seems to be minding the store; what's more, no one seems genuinely to care." By then each housing development in Chicago had an average backlog of a thousand unfilled repair requests. A survey of the 430 elevators across all CHA buildings revealed that 250 of them weren't operating. After a nine-month investigation, the FBI charged six CHA maintenance workers with pilfering millions of dollars' worth of paint, floor tiles, and roofing materials.

Byrne fired Swibel only after HUD threatened to withhold federal funding to the city if she retained him. In his place, she installed her former campaign manager. At the same time, she also expanded the CHA's board from five to seven members and appointed three white commissioners, changing the governing board from majority black to majority white. In 1983, after Byrne was voted out of office, her husband said he just so happened to bump into a pitcher from one of the Cabrini-Green Little League teams he coached.

"Hey, Mr. Jay, are you gonna be running the team?" the boy asked.

"No, Lefty. We got beat, ya know."

DOLORES WILSON

LIKE OTHER TENANTS at Cabrini-Green, Dolores Wilson watched warily as Byrne's time at 1150–1160 N. Sedgwick came and went. Her sons were harassed as they walked to work and returned home. Her two-door red Chevy, a gift from her son Michael, was towed and lost forever amid the mayor's wholesale cleanup. At work, her colleagues were so tuned in to Cabrini-Green that they'd ask her about every violent event that had been reported in the news. "Oh Dolores, are you okay? What was all that shooting about last night at Cabrini-

Green?" She hadn't heard any gunfire. If there was a shooting on Chicago Avenue, how could she hear it from her building a half mile north on Division Street? Now crimes committed anywhere on the Near North Side, two or three miles away, were being identified as occurring *near* Cabrini-Green. "If you stubbed your toe at Cabrini-Green, it was in the news," Dolores complained.

Her youngest brother was so spooked that he refused to visit her. "Don't you read the papers, Dolores?" he beseeched her. "They're talking about how many people are getting killed at Cabrini-Green." He worked at Mother's, a tavern with live music and a white clientele a short walk due east of 1230 N. Burling on Division and Dearborn Street. One night after his shift, Dolores's brother stepped out of the Gold Coast bar and three white guys jumped him, knocking out two of his teeth. It wasn't exactly funny to Dolores, but she definitely brought up the incident to her brother whenever she had the chance. "You said you're not going to visit me? I have my teeth. My family has their teeth. You're afraid to visit me because of what you *read* in the paper? Well, I'm not going to visit you from what I *see* happening to you."

Hubert Wilson was promoted from assistant head janitor in their building to head janitor. He was now on call twenty-four hours a day, and they moved from the fourteenth floor to a unit on the sixth floor. In case of an emergency, he had to be closer to everything. Their monthly rent—25 percent of their adjusted gross income—had been among the highest at Cabrini-Green. Most residents paid well under $100. But the new position included the benefit of a rent-free apartment. Dolores said she felt like a billionaire. They could keep their whole paychecks. She bought herself an extra pair of shoes and spruced up their apartment. "I had all my interior decorating going," Dolores said. The kitchen was impeccable, with a shiny stainless steel microwave, a yellow-tile backsplash, and a countertop lined with porcelain jars and decorative kettles. And they even had extra room. Che Che, who worked for Otis Elevator and was married with a child, bought a house from one of his coworkers on the South Side. Michael got a job replacing furniture upholstery and moved with his wife into the Cabrini rowhouses.

Although Hubert now had a staff of guys under him, he refused to sit at a table with a pen. He liked physical work, and he still woke early to pull the garbage and run the compactors. The other janitors called him "Old Man," even though he wasn't much older than many of them. One morning two weeks after Byrne's Easter festival, Hubert woke up with a bout of diarrhea. Dolores said she'd stay home to take care of him, but he shooed her off. He'd be fine. He didn't want to miss work. And since he was going to work, so should she. Then, like every day over the past thirty-five years, they kissed and traded I-love-yous. Early that afternoon he phoned the water department to tell Dolores he felt ill again. He was going home; he'd eat some crackers and take a nap. When Dolores returned to their apartment that evening, she saw he was sleeping. Quietly, she warmed up leftovers so food would be ready for him when he awoke. Her daughter-in-law stopped by, and Dolores sent her to the bedroom to check on Hubert. She came running back into the kitchen. "I think Daddy's dead," she cried.

When the paramedics arrived, they made everyone leave the bedroom. Dolores had put her fingers on the nerve in Hubert's neck; she thought she felt a pulse, but her own heart was pounding so hard it was difficult to tell. Hubert didn't look any different than he normally did as he slept. The paramedics finally came out of the bedroom, wheeling Hubert on a cart, a respirator in his mouth. Dolores stood to reach for his hand. Was he going to be all right? Could she ride with him in the ambulance? One of the paramedics pulled her aside. "I have to tell you that your husband is dead," she said. "We had to bring him out that way so people wouldn't snatch at the body and scream. You know how people do."

Dolores took a short bereavement leave from work. When she returned to the water department, she told her coworkers that she didn't believe what the doctors said about the diarrhea putting too much strain on Hubert's heart. She was pretty sure the paramedics killed her husband. It was an accusation delivered, as was her way, as a mouth-twisted aside, her voice remaining a trilling soprano. Her colleagues no doubt missed the simmering rage. "Oh, Dolores," they assured her, "no doctor would ever lie to you." But Dolores was if anything pragmatic. She didn't shut down because of a long list of perceived wrongs. She stayed busy at work and at home. She was a fifty-two-year-old widow, and in 1981 she had been living at Cabrini-Green almost half her life.

She was soon offered the chance to leave Cabrini. Her grandmother, the one who moved around for much of her young life, had settled down in Englewood on Chicago's South Side. She'd bought a corner house, with a coach house behind it where she let four men who'd been homeless live rent-free. When she passed away, the property went to Dolores's mother, and when she died Dolores's brother was named the executor. But he already had a house of his own, and their sister Connie had become an evangelist and given up on worldly possessions, selling off her furniture and quitting her job at the post office. The family assumed Dolores would take it. This was her opportunity to get out of public housing once and for all; she could own a place of her own. But she wasn't interested. She felt settled at Cabrini-Green. She told her brother, "I'm in the projects, but that's my home. I love my home just like you love your home." The property went to her grandmother's church, on the same street, and the church kicked out the squatters, demolished the two buildings, and turned it all into a parking lot.

PART TWO

CABRINI GREEN

HARLEM WATTS JACKSON

■

Cabrini-Green Rap

ANNIE RICKS

Aɴɴɪᴇ ʀɪᴄᴋꜱ ᴡᴀꜱ born on August 1, 1956, in Riverview, Alabama, the youngest of her mother's ten children. They lived in a one-room cabin, and Annie and her siblings arranged their narrow beds like jigsaw pieces around a potbelly stove. They had an outhouse, a slop jar, a well from which the children towed water for cooking and cleaning. Her mother took care of a white minister's family, earning $2 a day, working seven days a week. The Ricks children wore the hand-me-downs from the minister's flock. Riverview was a mill town on the banks of the Chattahoochee River, bordering Georgia. While Annie was growing up, there were separate schools for whites and blacks, separate parks, libraries, hospitals, and cemeteries. Annie would live in Chicago for almost fifty years and never once return to Alabama, even after Riverview and three other mill towns were incorporated to form a new city and their names and segregated schools were no more. "I told my own children they were blessed they didn't know about wells and snakes and the real prejudiced Jim Crow law," Ricks would say.

A constellation of freckles dotted the bridge of Annie's nose and reached her high cheekbones, which would sharpen to points each time she announced with a sly smile, "But I can't be mad." It was something she'd repeat her entire life, a calming mantra for whenever she felt adrift in the dark waters of a rage. She didn't know her father.

He'd left when Annie was six months old. At bring-your-dad-to-school days, she just said he was dead, which to her he was. "You know you're stubborn?" her teacher said. "Yes, ma'am." Her mother would tell her, "You're the strongest of my kids." When Annie was six, her uncle died, and the body was laid out in the front room of her aunt's house. Annie's sisters wept. One of them tore up her only pair of shoes on purpose to get out of going to the funeral. But Annie stared at her uncle without emotion until the casket was closed. She figured he had to be in a better place.

Annie's oldest brother was the first to leave for Chicago. When he settled into a job, he sent for their mother. Annie, being the baby, made the journey next. It was 1967, and a cousin and her husband were back in Alabama, visiting from Chicago. They let ten-year-old Annie ride with them on the return trip north. The cousin was also named Annie, and everyone in the family called her "Little Annie" and Ricks by her middle name, Jeffery. For the drive, Ricks put on her best dress and had her hair pressed. They rode through the night, reaching Chicago in an early-morning downpour, stopping first at Little Annie's South Side apartment. To Ricks, it looked like her cousin had struck it rich. The tidy apartment had a shiny stove and wood cabinets and an indoor bathroom with a sink, a flush toilet, and a bathtub that included a curtain and a shower. "Jeffery, you want to take a bath?" her cousin offered. "Yes, ma'am!" Little Annie turned on the tap, and Ricks watched gape-mouthed as steaming water poured out. She luxuriated in the hot bath, soaking and submerging and twisting for a full hour. Then her cousin fixed her a breakfast of eggs, toast, grits, and rice, her first meal as a Chicagoan. After she ate, Ricks asked if it might be possible for her to take another bath. The second one lasted longer than the first.

Within a year, the other Ricks children had followed Annie to Chicago. Their mother found a place for them in North Lawndale, a West Side neighborhood that had recently made the transition from a population of 100,000 whites to one of 100,000 blacks, a couple of generations' worth of shifting demographics time-lapsed into a few turns of the seasons. They lived on the top floor of a three-story walk-up in a three-bedroom apartment with a big kitchen, a living room, and a balcony. Ricks's mom would tell them, "Don't let the streetlights catch you." And Annie knew to be home each night before dark. But then she'd step onto the balcony and chat with friends whose parents were less confining in their rules. That's how Annie met Ernest Roger Bryant. His family lived in a similar walk-up a couple of doors down. She'd whistle, and Ernest would appear at his window. They talked for hours. He was two years older, and they started going together when Annie was twelve. He'd pluck at her globed afro, teasing that it wasn't her real hair.

"You know what happens when you lie down?" her mother asked one day. Annie said she knew. But her mother meant it less as a question than as a warning. And when Annie was fifteen and draping herself in her brother's oversize shirts, her mother knew what was up. Ricks gave birth to a boy, and they named him after Ernest. When she returned to tenth grade, a couple of months later, Annie's mother watched the baby during the day. In her senior year, Annie became pregnant with her and Ernest's second child, and left school for good.

Leap ahead a few years, to the tail end of the 1980s, and Annie and Ernest haven't traveled far from the West Side block where they met. They rented the second floor of a house just around the corner. Ricks was thirty-three, and she and Ernest now had five boys— Ernest, Shannon, Cornelius, Kenton, and Erskine (named for Inspector Lewis Erskine, a character from the TV show *The FBI*)—and three girls—Kenosha (whose name came not from the neighboring Wisconsin town but from the first initial of seven of Annie's closest friends), Latasha, and Earnestine. Rose, Deonta, Reggie, Raymond, and Raqkown would come later. The neighborhood had undergone its own changes. International Harvester, Sunbeam, the giant Hawthorne plant of Western Electric, Zenith, the Sears headquarters, and

many other manufacturers and retailers had shut their doors, tens of
thousands of jobs gone, with most of the smaller businesses making
an exit as well. Just one supermarket and one bank remained. The
neighborhood lost half its population, and it would halve again in the
coming decades. But Ricks had a job not far away in a factory mold-
ing plaster figurines. She enjoyed the work. "I enjoy everything that I
do," she liked to declare.

Annie believed that if she followed the rules, if she kept up her
part in whatever contract she signed, then she was entitled to all that
was promised her. Their landlord lived on the first floor, and when
the rent was due each month, she paid it. When something in their
unit broke, the landlord fixed it. That was how the arrangement was
supposed to work. So they all got along fine. One afternoon that fall,
Ricks saw the landlord arguing with a man in front of the two-flat.
She didn't know what about, but she figured later that it was that man
who torched their place. Annie's brother was staying with them, and
that night, while they were asleep, he was the first to smell the smoke
and yelled for everyone to wake up. They collected the children and
scrambled onto the sidewalk, managing to grab a few framed pictures
and toys on the way out. A half-asleep Latasha, hoping to curl up
in bed, was on her way back through the front door when somebody
saved her a second time that night. Then from the yard they watched
as their home and everything in it burned.

They tried staying at a cousin's apartment, and then another
cousin's place. Ricks bought groceries, took her younger relatives
with them to a diner. But even family could do only so much. She
came with a mob of kids, so many that they seemed to fill every space.
When they spent the night, they covered the floor like a carpet. You
couldn't put a foot down without stepping on one of them. Some nights
Ricks found herself herding her children along West Side streets with
no place to go. They walked a mile east, to Cook County Hospital, and
huddled in the lobby. Other families were there, too, and Ricks kept
watch as her children slept.

When Annie considered her predicament, the fact of it baffled her. She knew parents who were dope fiends and deadbeats, their children unkempt and uncared for. That wasn't her at all. She'd been providing since she was a teenager. Her children were well fed and neatly dressed, their hair combed and cut and braided. She made sure they finished their homework and were on time for school. She volunteered in their classrooms and attended their basketball games. When they pleaded for Air Jordans, she made sure they weren't the ones among their teammates to go without the shoes. So how could she not have a place for them to spend the night? The word describing her situation, that it applied to her, simply didn't make sense: *Annie Jeffery Ricks was homeless?* Without a permanent address, she lost her public aid. At her lowest, she asked a social worker at the public hospital to take her children, figuring they'd be better off in foster care, at least until she straightened out her housing mess. But the social worker refused. "Look how good these children are doing," the woman said. So Ricks divided up her family, parceling off the children to different relatives, keeping a couple here and a couple there.

One of Ricks's sisters had a job at Sears. A brother worked in a pickle factory. Two other brothers found service jobs at the University of Illinois at Chicago, and one of them was able to buy a bungalow in Marquette Park, where Martin Luther King had been attacked, and which was now well on its way to being largely black, Latino, and Arab. All Annie's siblings worked, and none of them, at that point, lived in public housing. But Annie was desperate and in need of what President Truman had described as every American's right to "a decent home and a suitable living environment." She applied for a CHA apartment. She put down as her first choice Lawndale Gardens, a development of 128 units in two-story rowhomes not far from her on the West Side. Her second choice was the Henry Horner Homes, a much larger complex on the Near West Side. Cabrini-Green was third.

On a December morning, amid a snowstorm, she let her third-oldest son, Cornelius, stay home from middle school. They were on the city's western edge, almost to the suburb of Oak Park, and Cabrini-Green was seven miles away. The CHA hadn't responded to her application yet, but Ricks had heard that there were openings at Cabrini. She told Cornelius they were going to get their apartment that day, and they were going to walk. In the snow. They started the journey by heading east on Harrison Street, in the direction of downtown with a hazy Sears Tower hovering in the distance. They passed churches and parks, a police station, and a checkerboard of overgrown empty lots now coated in white. They walked for an hour, and then another. Cornelius knew better than to complain to his mother about sore feet and cold hands. She sometimes walked on her own for entire evenings, and there was no dissuading her when her mind was set. They reached Ogden Avenue, the diagonal street designed to give white West Siders easy access to the lakefront. Ricks and her son walked it for another hour. They took Ogden as it spanned the Chicago River and continued over a valley of rail lines and factories. East of the river's North Branch, at the 1230 N. Burling building, the elevated roadway came to a blockaded dead end. They were still a mile from Lake Michigan. That's where Annie and Cornelius descended a flight of stairs. They'd made it to Cabrini-Green.

A woman at the housing office half-listened as Ricks detailed her plight, not seeming to care that they had trudged through the snow all the way from the West Side. The woman said no apartments were available—which was as much true as it wasn't. The Cabrini rowhouses were almost completely full. But a third of the red high-rise units were unoccupied and almost half of the white ones. The CHA hadn't fixed up and readied them for occupancy. The agency said it couldn't afford to do the repairs to rent the empty units, and not just at Cabrini-Green but at public housing all over the city. Ricks dismissed the woman's words with a wave of her hand. She didn't want to hear about budgets or the agency's problems. To her the math was simple: twelve hundred apartments at Cabrini-Green with no one in them, and she and her children burned out of their West Side home, wandering the streets, sleeping in a hospital lobby, needing just one. "No, ma'am," she said. She motioned for Cornelius to take a seat. They weren't going anywhere. Ricks could pester, persist, accuse, as if on a loop. She announced that she was about to raise some hell. She was going to call all the TV news networks, channels 2, 5, 7, and 9. She started naming the journalists who'd be interviewing her— Walter Jacobson, Ron Magers, Oprah. "Why lie and say you don't have an apartment for my family? I know you have an apartment. There are so many apartments here." Ricks cut herself short, a look of surprise suddenly giving way to a self-preserving grin. "But I can't ever be mad."

By then, though, the woman at the desk was relenting. She sent Ricks to one of the white high-rises, next door to Dolores Wilson's building. The tower at 660 W. Division was a fifteen-story plain box, a giant filing cabinet with a facade the color of cigarette-stained teeth. The elevators were out of order and the stairwells were dark. The fifth-floor apartment Ricks entered looked like a crypt. Plywood covered the windows. Trash and old clothes were clustered along the floorboards like blown leaves. The kitchen cabinets dangled or were missing altogether. Ricks surveyed the run-down surroundings, counting four bedrooms. There was a full bathroom on one side of the unit and a half bath on the other. The front room was large enough for a dining table and a sofa, and it was connected to the kitchen, which (she checked) had a working stove and refrigerator. The ceilings were high, the walls made of seemingly indestructible cinder block. She smiled. What Annie Ricks saw looked like a home.

RICKS KNEW VERY little about Cabrini-Green or its reputation when she moved in with her family. She hadn't watched *Good Times* or *Cooley High*. She wasn't living there yet, in 1983, when a documentary on the Jesse White Tumblers, *The Ambassadors of Cabrini*, aired on TV, contrasting the development's perilous streets and fetid stairwells with the militarized grace of these athletes leaping and flipping and somersaulting. "Most of the young little white kids, they grow up, their mother can send them to a school, or they can take up gymnastics, use all the apparatus, or go to swimming. So they can take up ballet," a young Jesse White Tumbler named Marcus says in the film. "We don't have money to do that. If we had what they had, we'd be equal. But we can't be equal without nothing to help us out." Ricks would soon get to know many of the two hundred Cabrini residents who were interviewed for the Free Street Theatre's musical *Project!*, and she'd meet Dolores Wilson, who served on the Free Street Theatre's board. *Project!*, from 1985, mixed sketches and original songs with testimonials from Cabrini-Green tenants displayed on seventy monitors stacked on the stage to look like the red and white highrises. An actor rapped, "Cabrini means red, and Green is white / And if you want to stay alive, you better get that right!"

Ricks hadn't seen the recurring *Saturday Night Live* skit from around the same time featuring a teenage single mother named Cabrini Green Harlem Watts Jackson. Especially after Mayor Byrne's stay, the housing development had entered the pantheon of proper names of scariest black places in America. The character, played by Danitra Vance, the show's only black female cast member, wore pigtail braids and a T-shirt tucked into a miniskirt. "I was at home, my mama was fixing some cornbread, black-eyed peas, candied yams with neck bones, and some Kraft macaroni and cheese," Cabrini Green Harlem Watts Jackson relates. "I said, 'Hi, Mama. You wanna hear a joke? I'm pregnant.' She said, 'How did that happen?' I said, 'How am I supposed to know how that happened? You never told me nothing about things like that, the school don't teach us about things like that, you're asking me how that happened?'" The week Oprah Winfrey traveled from Chicago to guest-host *Saturday Night Live*, she played both the slave of *SNL* producer Lorne Michaels and the mother of Cabrini Green Harlem Watts Jackson.

Ricks did have a brother-in-law who worked at the giant Montgomery Ward warehouse behind the Cabrini rowhouses. He told Annie that people at Cabrini-Green were shooting. But that didn't mean anything to her. People were shooting on the West Side. People were shooting on the South Side. "Because I hadn't heard of Cabrini-Green, I came to it fresh," she would say. Ricks wasn't in her new apartment twenty-four hours, though, before two police officers banged on the door. Apart from a few garbage bags filled with clothes, the four-bedroom unit was bare. Everything else they'd owned had burned in the fire. Annie was lying on the floor, napping through the riot of noise created by her children, when the police entered. "Is this your apartment?" a lady cop demanded, assuming they were squatters. Boards still covered the windows. Annie handed over her lease. "All these children in here and no furniture," the officer said. "I could call Department of Children and Family Services on you." For the cop, Ricks was the embodiment of the Cabrini-Green image, Danitra Vance's character in the flesh. For Ricks, the threat seemed *almost* comical.

"I don't care what you call on me," she said. "I tried to get my children to DCFS. They wouldn't take them. These kids are well-dressed and taken care of." She snickered. At least she now had an apartment and was in the process of getting on public aid. She had to show the paperwork for that, too. Ricks used a voucher to pick out furniture at a store nearby that catered to families from Cabrini-Green. She got bunk beds for the children, a kitchen table, and chairs. The cabinets were reattached, the boards removed from the windows. She painted the front room blue, the kitchen yellow, and the girls' bedroom pink. She laid tile in the kitchen. Annie's mother moved in with them, watching the children when Ricks worked, but Ernest didn't stay there, at least not officially. He and Annie hadn't married, and she didn't want to be disqualified from public aid. "If I had put him on my lease, my rent would have been sky-high," Annie said. The lady cop returned to check on Ricks, and she saw the kitchen set and the beds, the new paint and the light coming in through the windows. She nodded with approval. But that wasn't good enough for Annie. "Why are you going to say you're calling DCFS on me?" she asked. "You're a mother, too. Why you going to say that to me?"

Unlike Ricks, Kelvin Cannon was well versed in Cabrini-Green lore. As a lifelong resident and as a young man, he took a perverse pride in the cartoon depiction of his home. If the media helped form its iconic image as a world of utter deprivation and unrelenting violence, then saying you were from Cabrini projected a kind of power. You mugged into the fun house mirror. You fed people the stories they wanted to hear about the murders, the gunfire, the beatings, the people lined up outside the high-rises to buy drugs. After Mayor Byrne added to its infamy, the name Cabrini-Green was evoked to disparage anything as derelict or dangerous, from neighborhoods in other cities to a temporarily out-of-service elevator in a high-end apartment building. Cabrini-Green became one of the stand-ins for the city itself, something people shouted in recognition when they heard the word "Chicago," like Al Capone or Michael Jordan. A resident says in *Project!*, "It's just automatic, when you say 'Cabrini-Green' people start freaking out. It's almost like a state of mind."

When Cannon was a teenager, he'd be in lockup with hardened gang members from other parts of Chicago. A news story about his home would come on the television, and guys would treat him with respect, asking him if he had ever been inside the Rock, or if it was true about the barrage of rifle fire on New Year's Eve. "Yeah, I've been over there," a man in the jail said to Cannon, as if describing a close call on a military tour. "I'm never going back."

On the set of *Good Times*, at CBS Television City in Los Angeles, the black cast members and the white producers argued about the portrayal of the show's Cabrini-Green family. They debated authenticity and responsibility, which led to scouting trips to Chicago's Near North Side. Esther Rolle, who played the Evans family matriarch, demanded that her character be married, the husband hardworking and present in his children's lives. She and the other actors insisted that the sitcom grapple with real social issues—gangs, welfare, racial discrimination in housing and employment. They wanted the episodes to focus more on the youngest Evans child, Michael, and his middle school protestations of Black Power. But the producers wanted more lines for J. J.—James Jr.—the eldest child, an out-of-work artist living at home, played by the comedian Jimmie Walker. Walker delivered his tagline as if possessed, slapping his hands, throwing back his head and holding the pose—"Dyn-O-MITE!" It was no contest. Jokes about the Nation of Islam lost out to Walker, with his jelly-limbed physical comedy and exaggerated minstrelsy. "I wanted to make my favorite sandwich today, peanut butter and jelly," J. J. announces. "But there wasn't no peanut butter, and there wasn't no jelly. So I was forced to make a ghetto jam sandwich. Two pieces of white bread JAMMED together."

When Cannon saw *Cooley High*, at the McVickers, Cabrini residents packed the downtown theater. They pointed out the familiar locations and the locals who'd been hired for small parts. "It was an inspiration," Cannon recalled. The movie's main character, Preach, convinces a girl to go on a date with him, and the two of them stroll together over the Ogden Avenue Bridge—*Kelvin's bridge!* Cannon would attend Cooley High himself. But the film was set in the prelapsarian times before the King riots and cop killings. The characters are members of the Cooley class of 1964, back when Kelvin was a baby. Preach, Cochise, and Pooter, wearing their cardigans and caps, look almost like they're out of *Archie Comics*. They skip school to clown at the nearby Lincoln Park Zoo. They meet up at a local soda shop, and join two gangbangers—played by real Cabrini gangbangers—to joyride through the Loop. The movie seemed like an artifact, a portrait of all that had changed in a brief decade, in both the reality and the perception of Cabrini-Green. The teens in the film aren't afraid that crime is their only option; they crave something more stimulating than the drudgery of low-skilled factory jobs that are still there in abundance.

Cannon grew up alongside Demetrius Cantrell—Sugar Ray Dinke—who made one of the first rap songs ever to come out of Chicago. The night in 1981 when the singer Larry Potts was murdered in the 1230 N. Burling rec room, that was the end of the Electric Force Band. No way could they play after that. But Dinke continued to make music on his own. He'd rap about anything, verses streaming out of him about news events and sports figures and bits of advice. Oprah Winfrey visited Cabrini-Green in the early eighties to feature the Jesse White Tumblers. The fascination with crime and degradation there meant that the media focused also on the successes that seemed to blossom from the cracked asphalt. There were multiple news stories about Anthony Watson, who grew up in the rowhouses and became a navy commander in charge of a nuclear submarine ("Rough Sailing, But He Beats Cabrini Odds") and stories of teachers, tutoring programs, and entrepreneurs. As Oprah interviewed Jesse White outside Schiller school, the gym teacher interrupted her. "Hey, Dinke," he shouted as the young man passed by. "Come do that rap about Cabrini-Green."

Dinke did it on air, a cappella, a rap about the stench in the stairwells and the body he saw on the blacktop and the time someone fired two shotgun blasts through the front door of his apartment and what it was like to perform in a band there. A music producer watching wanted to record the song. He didn't know anyone at Cabrini-Green, so he looked in the phone book and called a neighborhood liquor store. He asked if they knew about the kid who rapped on *Oprah*. Of course they did. They passed along the message, and Dinke went to a recording studio and did the song in a single take.

"Cabrini Green Rap," like other straightforward raps of the mid-eighties, lands hard on the AA BB end rhymes, marching ahead to the steady beat of synthesized drums. But the song is alive with the tragedy of a specific place that Dinke loves. In a thumbnail history, he rhymes, "Cabrini-Green was built for the low income / So the people could live a little better than bums." He relates how the lawns and flowers disappeared, how the development deteriorated and the name itself came to evoke an idea about cities in decline. Most of all he conveys the particulars of a friend's murder, showing how violence remains singular and devastating even when it's accepted as commonplace: "I never will forget my man Larry Potts / Or the terrifying night that he got shot."

Concentration Effects

KELVIN CANNON

JUST AS BO John had predicted when recruiting Kelvin Cannon and the other middle schoolers behind Schiller, the Gangster Disciples took over much of Cabrini-Green. The GDs presided in the white high-rises, the rowhouses, and all but a handful of the red towers. Kelvin grew with the gang. He was dedicated, and he paid attention. He studied what he needed to do to be a leader. "To be aware is to be alive," he'd say. Standing outside 714 W. Division, he'd notice a car creeping past, the passengers sneaking looks at him and his friends. He'd shepherd everyone inside, and they'd already be on an upper floor when the car turned up Division to ride up on them. They called him "Righteous Folks." He didn't present himself as a wild gangbanger. He didn't hurt people for no reason or sneak up with guns on guys. He was someone who'd take up a collection when a fellow Disciple was hurt or killed.

Cannon was on the Cooley High baseball team, pitching, catching, and playing left field. After a brawl at the school, he was called into the office. He claimed innocence, challenging a police officer to search him for weapons. He'd forgotten about the reefer he had hidden inside his hat, and he was expelled. Cannon had broken a hundred rules, for fighting and weapons and drugs. But he'd never before been caught like that, so the punishment seemed unjust. Still, he had to accept that he'd slipped up. "What goes around comes around," he reasoned.

When Cannon was sixteen, two of his friends told him they were going to rob an older woman who was selling drugs out of her third-floor apartment in Dolores Wilson's building. The woman had moved to 1230 N. Burling from the South Side, and she and her sons were believed to be Blackstone Rangers. "They were like opposition to our organization," Cannon said. "Another gang comes into your territory, tries to set up shop, you know the repercussions. You got to shut them down." He and his friends entered the next-door building. Cannon hid on the ramp as the other two knocked on the woman's door, saying they were there to buy weed. When the door opened, they pulled out guns and called for Cannon to join them. He saw three women seated on a couch, all of them his mother's age. Rifling through the kitchen and bedrooms, he bagged up whatever drugs and money he found, and they left. Except for a shot fired that didn't hit anyone, the robbery was unremarkable, one incident among many. Cannon might not have remembered it if it hadn't come back on him.

A couple of weeks later, he was at the county jail, on an unrelated gun possession charge, when he saw his cousin Greg across the bullpen. Greg, it turned out, was in there for the robbery in the Burling building. They were first cousins, and someone had fingered Greg for the crime, mistaking the two of them. When Greg realized the mix-up, he told the police. In an interrogation room, a cop said to Cannon, "We got your cousin. He said you could clear him and tell us who else was involved." Cannon couldn't inform on his friends—that would be a violation of the Disciples code. Greg shouldn't have said anything, either. "That's not being an honorable Disciple," Cannon would say. But he couldn't hold a grudge against family. Greg later went to the penitentiary on a thirty-year sentence for ten robberies he committed in Chicago and Michigan. Cannon tried to fight his own case in court. By the time a judge ruled on it, nearly two years later, Jane Byrne was living at Cabrini-Green and the city was making an example of anyone arrested there. Cannon was eighteen then, a father, and charged as an adult with home invasion and armed robbery.

It was his first conviction, and he was given seven years, of which he'd have to serve at least three and a half.

At Stateville, a bunch of guys from Cabrini-Green worked in the kitchen, and they helped Cannon land a job there as well. As he delivered food, he got to know the Disciples leadership, the board members, and they took note of how Cannon carried himself. He was mature and smart, keeping himself busy, never boasting about what he'd done or who he knew. They pointed out to Cannon that he had an out date. Many of them didn't, and they told him he had no business in prison, that he needed to go home and take care of his family, raise his newborn son. Cannon's father would visit, and they'd pray together that Cannon would come to lead a truly righteous life. "Prison made me a better person," Cannon would say. "I was so wild back then. I had no regard for other people. I was going to get killed or be in the penitentiary the rest of my life. You listen to the old-timers, as a daily thing, how they lost their life in prison. They were trying to save me."

Cannon became friends with Johnny Veal, who was in his second decade of a 100- to 199-year sentence for his part in the 1970 slaying of Sergeant James Severin and Officer Anthony Rizzato. Veal was Cobra Stones, but he told Cannon that inside the prison they both rode Cabrini. The black community from the Near North Side could feel like a small town, everyone bunched together on an island of a few blocks and parceled into a handful of neighborhood schools. All around them were wealthier and whiter communities, cutting them off from the black neighborhoods to the south and west. Everyone knew one another's families, if not each other personally—*You're a McNeal, right? A Brown? One of the Campbells? Them Marlow's people.* Veal explained that this was also the meaning of Cabrini-Green. They were from the same village, and the guys in the joint from Cabrini, whether they were Disciples, Cobra Stones, or Vice Lords, represented their neighborhood. "He taught me never to forget my roots," Cannon remembered. "He was schooling me, telling me not to tear down Cabrini-Green but to help build it up."

■ ■ ■

EVEN WHEN THE Cabrini-Green high-rises were new, in the 1950s and 1960s, the first residents already talked about their home as being doomed. "We're living on a gold mine," Hubert Wilson had always told his family. Cabrini was a mistake, and the city wouldn't allow a black settlement to remain so close to downtown and the Gold Coast. People spoke of the secret plans hatched by Realtors and government officials, seeing signs of their forced removal all around them. Then in 1973, the city, along with downtown real estate and business interests, published "Chicago 21," a comprehensive development plan to reverse two decades of economic decline and flight to the suburbs. "The Central Communities must be revitalized to again become desirable both for living and for working," the report stated. "They must be efficient, economic and secure, and they must also provide maximum opportunity for human fulfillment." Eleven centrally located communities, including Cabrini-Green, were targeted for reinvention and resettlement. To survive, the city had to rebuild its tax base. The poor, black, and brown people who currently sought human fulfillment in these areas believed that their homes would be taken from them, remade, and given to wealthier newcomers who were themselves considered more desirable.

The Cabrini-Green activist Marion Stamps helped form the Coalition to Stop Chicago 21, and the citywide alliance of blacks, Latinos, and working-class whites was one of the groups that got behind the first mayoral run of Harold Washington, in 1979, and then organized for him more systematically during the 1983 election. Washington had been a precinct captain on the South Side, a position passed down to him from his father. He rose up in Daley's Democratic machine, serving in the Illinois House and Senate, and he won a seat in the US House of Representatives in 1980. But Washington showed flashes of independence that put him at odds with the party stalwarts. He split with Daley over police abuse in the black community, demanding that the city create an independent review board to investigate cases of brutality and misconduct. A preternaturally gifted orator, Washington could sound like he was performing Shakespeare, a Falstaff with a large, round face creasing into a smile, even as he was telling an opponent to drop dead. "The term 'patronage,' like its close cousin, 'paternalism,' comes from the Latin 'pater,' meaning father," he announced to a crowd in 1980. "Every citizen pays for the patronage system in poor city services and poorly constructed facilities. We pay for it when city government puts big business and downtown interests before the needs of average citizens. But my community, black Chicago, suffers most deeply."

In the summer of 1982, African Americans boycotted a Grant Park music festival called ChicagoFest, after Mayor Byrne cut black representation on various city boards and commissions. Coming near the end of Byrne's first term, the protest was a surprising success. Stevie Wonder, set to perform, cancelled, as did dozens of other acts. Suddenly, there was a coordinated effort to sign up unregistered black voters in the city. People were registered at libraries and churches, at welfare offices and grocery stores and in public housing. Marion Stamps helped lead the voter drive on the Near North Side, and Dolores Wilson walked her high-rise, signing up her neighbors. Citywide, more than 130,000 people were added to the rolls, with most of them coming from predominantly black precincts. "I'm running to end Jane Byrne's four-year effort to further institutionalize racial discrimination in this great city," Washington announced, folding Byrne into his critiques of President Reagan's urban austerity measures. During the speech in which he declared his candidacy, Stamps interrupted Washington, yelling, "Harold, you are like the Second Coming."

Stamps ran for alderman of her ward, and when Washington campaigned at Cabrini-Green alongside her, thousands of people flocked around them. Ed Vrdolyak, a councilman from the city's industrial Southeast Side, tried to motivate Byrne's supporters by feeding on their dread of a racial takeover. African Americans now made up 40 percent of Chicago's population. "It's a racial thing," Vrdolyak said.

"I'm calling on you to save your city, to save your precinct. We're fighting to keep the city the way it is." Only a handful of white politicians and very few of the city's media outlets endorsed Washington. When asked whether her support of a black candidate wasn't the flip side of the same Chicago racism, Stamps rejected the idea. "They have tried to keep us for so long not being proud of our own," she said. "And I'm saying, 'Feel good about yourself and vote for one that is part of you.' Harold Washington is blood of our blood. We don't have to apologize about that. He is a father of our fathers. And a son of our sons."

Stamps lost her bid to the neighborhood's longtime incumbent, Burt Natarus. But with a record turnout in black precincts, Washington won the Democratic primary, as Byrne and Richard M. Daley, the Cook County state's attorney and son of the former mayor, effectively split the white vote. Daley endorsed Washington in the general election, even as many whites in Democratic Chicago chose race over party. Washington's Republican opponent campaigned with the slogan, "Epton for Mayor—Before It's Too Late." Out of a total of 1.3 million votes cast, Washington won by fewer than 50,000, becoming the city's first black mayor. The attacks on his administration began immediately. Vrdolyak's coalition of twenty-nine white aldermen halted almost all activity in the fifty-member city council. Washington was able to cut the budget by trimming the number of city workers, and he signed on to the Shakman decrees, which made it illegal to hire or fire city employees based on political expediency. But most of his attempts at reforms, or even at appointing board members to the various city departments, were blocked. Every council meeting was a battle of insults and innuendo. Not all the "Vrdolyak 29" were outright racists, but some were. "Scurrilous hooligans," the grandiloquent mayor called his foes. The stalemate came to be known as the "Council Wars."

Under a black mayor, with a black head of the CHA and a black police chief, troubles in public housing persisted. The CHA was crippled by debt. During the Reagan years, the HUD budget was cut by 75 percent, while money from rents continued to dry up, and the CHA desperately needed to upgrade its aging and poorly maintained properties. At Cabrini-Green, bedroom ceilings leaked, sinks and tubs didn't work, windows wouldn't close, and apartments lacked refrigerators or stoves. The CHA was already performing a kind of triage. It spent almost a tenth of its annual budget of $146 million on elevator repairs alone. Another $44 million was needed for asbestos removal, and three-quarters of a billion dollars over five years for the repair of its 1,300 buildings.

Washington picked Renault Robinson to right the tottering agency. Robinson had helped found the Afro-American Patrolmen's League, a group created to challenge racism within the Chicago police department's own ranks. But as the chairman of the CHA, he warred with the agency's executive director, leading to deadlock. He fired dozens of elevator repairmen after an investigative news team showed them loafing on the job but did so before finding replacements; residents in the high-rises were relegated to the stairs. A tenant on her way to the hospital had to walk down ten flights— she collapsed and died. Robinson abruptly dismissed 260 plumbers, pipe fitters, glaziers, and electricians for not doing "a full day's work for a full day's pay." With an understaffed maintenance crew and amid a brutal winter, however, the boilers broke, buildings went without heat, and pipes froze and burst. The CHA was forced to rehire the craftsworkers and pay them $400,000 in damages. Marion Stamps called for a rent strike at Cabrini-Green, demanding that the Washington administration do something about the insufferable conditions. By Washington's third year in office, his housing agency was running an $8 million annual deficit. The CHA lost out on $7 million in federal funds because it missed the grant deadline. "The CHA was systematically ignored and raped," Washington told a hundred residents at 1230 N. Burling. He listed the mayors who preceded him, saying that they "didn't fix the elevators, they didn't fix the trash chutes, they didn't give you police protection. They didn't give you a damn thing." Washington said, "I didn't build the CHA. . . .

Had I been mayor twenty-five years ago, this mess would never have been created."

The idea of tearing down the high-rises, previously considered politically and logistically untenable, began to be floated as a thought experiment with greater frequency. The chief financial officer at the CHA mentioned to reporters that the agency might have to sell off its well-located properties just to pay its bills. Renault Robinson let it be known that developers called him frequently to express interest in buying Cabrini-Green. One group of Realtors took him out to lunch and offered $100 million for a clear title to the Near North Side development. "The sale would remove what has become one of the most crime-plagued, socially unacceptable public housing projects in the country. And it would pave the way for one of the most positive residential real estate developments in the city," said Alderman Ed Burke, the other ringleader of the Vrdolyak 29. Alderman Natarus, an ally of Washington's, sent the mayor a worn copy of a letter he'd written to Richard J. Daley in 1972 that opened with the line, "As you know the situation at Cabrini Green continues to be precarious." He began his letter to Washington the same way, emphasizing that little had changed in more than a decade. "The repeated incidents of sniper fire have occurred in every administration in which I have served as a public official," he wrote. He proposed a possible solution. "It is not inconceivable that three story walk-up units could be built within the open spaces of Cabrini Green and that as soon as these units are completed for occupancy, steps could be initiated to demolish the high-rise buildings."

The suggestions resulted as well from the dramatic changes to the areas surrounding Cabrini-Green. By the 1980s, the revitalization of the central communities detailed in the Chicago 21 plan had started to come to pass. The anti-urban impulse that had sent middle-class families fleeing to the suburbs had reversed itself, and young professionals—the decade's Yuppies—wanted to live not in the all-residential bedroom communities of their parents but in city centers near their jobs and one another. One of the first new developments to be built alongside Cabrini-Green was Atrium Village. The property bordering the El tracks there had been vacant for almost a decade, and the city's Department of Urban Renewal was happy to give it away. In 1978, an alliance of four local churches partnered with a developer to build a 307-unit complex of townhomes and geodesic high-rises, each of them facing inward and creating a kind of fortress. Atrium Village was a racially diverse, mixed-income development, with quotas for the number of renters receiving heavy federal subsidies. Jesse White moved into an apartment there, and Dolores Wilson served on Atrium's board. The churches conceived of Atrium Village as a bridge between Cabrini and the Gold Coast, a step up out of public housing. But more people ventured across the bridge from the other side. The market-rate units were snapped up, ushering in a building boom.

From 1979 to 1986, the communities around Cabrini-Green experienced $900 million in new construction or upgrades. Just to the south of the rowhouses, on the other side of Chicago Avenue, fluttering banners proclaimed a new creative community called River North, chock-full of renovated lofts, restaurants, and office space. To the north, townhouses along Larrabee now fringed the Cabrini development. In 1979, Waller High was transformed into the new Lincoln Park High School, with a revised curriculum, a refurbished building, and money for a band and a chorus. The principal was permitted to recruit high-performing students from out of district, and the student body fell from over 90 percent black to 50 percent. At the same time that the *Tribune* was publishing a twenty-nine-part series it titled "The American Millstone," focused on "an underclass that is mostly black and poor and hopelessly trapped in the urban centers of Chicago," the city was also being celebrated for its "urban renaissance." A 1986 feature on the Near North Side noted that "new boutiques and rehabbed apartments are closing in on one of the country's most spectacular examples of failed public housing:

Cabrini-Green." Cabrini was becoming an island of black abject poverty amid a sea of encroaching white affluence.

As the black population of the Near North Side dwindled, those remaining were also in the process of being reconceptualized. Welfare families, single mothers, the perennially unemployed—the social programs of the Great Society had failed to lift them out of the depths of generational poverty. William Julius Wilson, then a sociologist at the University of Chicago, explained that larger structural transformations had robbed black neighborhoods in Chicago of much of their remaining wealth and stability. From 1967 to 1987, amid the shift from an industrial to a service economy, the city lost some 325,000 manufacturing jobs. Two-thirds of the job growth in the Chicago area occurred outside the city. Middle-class and working-class African Americans moved from their longtime neighborhoods, following jobs and better housing opportunities elsewhere, leaving behind a higher concentration of poor families. During the 1970s, the percentage of families living well below the poverty line increased in twenty-five of Chicago's predominantly black communities. The number of black communities with unemployment rates above 15 percent jumped from one to fifteen. In his book *The Truly Disadvantaged*, Wilson argued that those who were financially better off had acted as role models for the poor, buffers against both economic and personal collapse. Their departure, he asserted, compounded the impact of joblessness, resulting in "concentration effects"—the "social pathologies" and "ghetto specific culture and behavior" that came to dominate. By 1983, three-quarters of the black children born in Chicago were to unmarried mothers. Graduation rates at public high schools in black and Latino neighborhoods fell below 40 percent. The average age at which African Americans were arrested for violent crimes fell lower and lower. "Welfare and the underground economy are not only increasingly relied upon," Wilson wrote, "they come to be seen as a way of life."

Wilson was building on the controversial Moynihan Report of

1965. Daniel Patrick Moynihan, then the assistant secretary of labor under President Johnson, wrote of the "tangle of pathology" afflicting families in black ghettoes. His language of disease distracted from his warnings about the increasing breakdown of social structures in impoverished black neighborhoods. Wilson, an African American scholar, was stressing that the problems of out-of-wedlock births, single-mother families, lack of formal education, and criminality had only worsened in the ensuing two decades. People in these high-poverty areas had fallen further outside the occupational and behavioral mainstream, and more than ever, he insisted, these issues needed to be addressed candidly. During the Reagan years, Wilson was sometimes mistaken for a conservative, since his rejection of liberal orthodoxies intersected with a view of inner-city ghettoes fostered by scholars on the right. Charles Murray, among others, denounced welfare programs as the actual cause of the black family's dissolution, contending that poor people made the rational market decision to forego low-wage work for government handouts, and that parents even calculated to have more children out of wedlock for the bump in benefits another dependent provided. By this reasoning, it was more charitable to do away with all social welfare. But Wilson debunked these fallacies; the earned income tax credit, for instance, was established in the 1970s, and by conservatives' logic, the expanded benefits for those who worked should have driven down the welfare rolls. Like Moynihan, he proposed not that government assistance be slashed but rather doled out more generously and strategically. Still, the popularizing of the notion that neighborhoods of concentrated poverty were destroying their inhabitants' lives and creating a permanent "ghetto underclass" had a similar effect on many moderates and liberals at the end of the century as it had on the CHA do-gooders in 1950 who set out to raze the Little Hell slum—they were emboldened to rethink the government's role in these communities, which now meant talking openly about retrenching the welfare state.

Nowhere in Chicago were the concentration effects that Wilson lamented more pronounced than in the forced isolation and density of high-rise public housing. By the eighties, Chicago was home not only to three of the country's twelve richest communities but also, amazingly, to ten of the country's sixteen poorest census tracts, all of them containing large public housing complexes. Wilson pointed out that 83 percent of those living at Cabrini-Green in the early eighties were on welfare, and of the families with children, 90 percent of them, at least officially, had only a mother at home. "The projects simply magnify these problems," Wilson wrote. The Chicago Urban League issued a report calling for an end to the tyranny of the high-rises. Editorial boards announced that Chicago was ready to rid its landscape of these civic disgraces.

For residents of Cabrini-Green, the public discourse about their fate was more evidence that the mass evictions they'd always feared were finally upon them. Yes, they lived in what was now being rec-ognized as concentrated poverty. But by necessity they'd also made it their home. Families grew up next to one another, generations of them. They watched one another's children, shopped together, shared food, stepped up when a family lost a loved one or was in need. Many relied on an off-the-books economy of hairdressers, handy-men, makeup artists, babysitters, auto mechanics, manicurists, shoe shiners, tailors, cooks, carpenters, and candy sellers. People bar-tered services and passed along news of job openings. Theories about concentrated poverty often ignored all that. Cabrini residents didn't know how they'd manage elsewhere without the support of these longtime networks. They worried that crime and poverty would exist wherever they were sent. And it seemed doubtful that tearing down the most visible incarnation of these concentration effects would do away with the discrimination, government inefficiency, and fear that went into making the Cabrini-Greens of the city and the country in the first place.

A group calling itself the Concerned Tenants of Cabrini Green Homes announced a community-wide meeting to warn of the power brokers who were then plotting against them. The flyers that were taped to lampposts and the walls of high-rise lobbies asked, "Are we next? Where will we go??" In the handbill's line drawings, men in suits identified as "Investors," "Developers," and "Land Grabbers" stand to one side, laughing. A man labeled "Tenant" cowers, while an extended leg with "CHA" written along it delivers the banishing kick.

KELVIN CANNON

ON JUNE 7, 1984, Robert Cannon drove to Stateville to fetch his son. Kelvin had been barely eighteen when he was sent away; now he was twenty-one, with a body reshaped by weight lifting and starchy prison food. His waist size had stretched from a twenty-eight to a thirty-six, his neck thickened, and his chest now rolled with heavy muscles. He didn't own a single item of clothing that fit. Cannon had earned $800 working in Stateville's kitchen, and his father took him straight to Maxwell Street, the freewheeling outdoor market, where Cannon bought everything from pants and shirts to socks and underwear. His next stop was Cabrini-Green.

In the months before his release, Cannon needed to find a place where he could live. When he left for prison, his son was three months old. The boy was now four and a half, but his mother had married and had additional children. He obviously wasn't going to stay with her. Cannon's mother lost the apartment in 714 W. Division because of his conviction, and his father didn't invite him to join his new family on the South Side. Cannon reached out to William Blackmon, his best friend growing up. William was home after a stint in jail, and he told Cannon how good it felt to be back on the street. Cannon said he'd soon join him. "Just lay low until I get out there," Cannon said. But then word reached him in Stateville that his friend was killed while trying to break into a Cabrini apartment. Another childhood friend now lived in 1230 N. Burling—the same building in which Cannon committed the robbery that got him locked up—and the friend said

Cannon could stay with him. So that's where he went, another former inmate paroled to Cabrini-Green.

Cannon had heard countless times that prison either makes or breaks you. It definitely didn't break him. He'd served time with the leaders of the Disciples and won their trust, proving himself over the years. He'd gone in strong, and come out stronger. Now, back home, people looked up to him for the time he'd served. He was a role model. The Gangster Disciples were organized, with regents, assistant coordinators, coordinators, governors, and a system for levying taxes. Cannon was named a Disciples don at Cabrini-Green, and a year later, at twenty-two, was appointed governor of all Cabrini-Green. "Cannon would be a success no matter what he did," a police officer who worked the Cabrini-Green beat said. "If he grew up in suburban Glencoe, he'd be a doctor with the prettiest wife and the biggest house. At Cabrini-Green, he aspired to be like the most successful people in his community. He hit it out of the park. He became that guy."

Cannon didn't have to recruit, since most of the boys wanted to be Gangster Disciples. It looked like fun to the young guys, and it paid for diapers and dinners, sneakers and jackets. Cannon told the ones who were thirteen or fourteen—the same age he'd been when Bo John recruited him—that they needed to spend more time in school before they could choose to be a GD. They had to have some education first. He liked to think of himself as a peacetime leader, not a war one. He'd practice drawing his gun in his bathroom mirror, working on his speed like in the Westerns. One time he accidentally shot a hole in the medicine cabinet. But he wasn't into battling with the other organizations. He'd get up in the mornings, walk the land, see what was going on, and learn what had happened the day before. Sometimes he had to have friends violated, physically punished, when they'd snatched a purse or robbed another member's house or raped. There was a code, and it had to be enforced. A guy might get GD glasses—punched in the face so many times his eyes swelled up into puffy rolls. Or they could get worse.

Cannon's goal was to smooth out problems before they reached that point. He thought some of the unity he'd seen in prison could be used back home. Guys at Cabrini had gone to schools together, slept in the same apartments, eaten at the same tables. And now that they were older and part of different organizations, he didn't see why they couldn't still coexist. He threw back-to-school picnics and organized sports leagues. The smoother he could make each day, the less chaos there was, the better he felt he was doing his job.

As governor, Cannon never got rich. This wasn't *Scarface* or *Miami Vice*—no boats or sports cars or penthouse apartments along the Chicago River. He even held down legit jobs at the same time, driving out to the suburbs to work as a forklift operator at a plant that mass-produced prepared foods. Just as William Julius Wilson had documented, most of the low-skill jobs had moved out to the suburbs, if not out of the Chicago area and the country altogether. You couldn't walk or take a bus to a suburban office park or manufacturing plant. You needed a car, and one that didn't break down in winter. Which is what happened to Cannon. His car stopped working, and he lost the factory job.

Cannon did get his own apartment in 1230 N. Burling, and he managed to fix it up. Outside, the hallways were scribbled with tags. Inside his thirteenth-floor apartment, everything was bright rugs and blinds and plush furniture. He topped glass tables with framed photographs and hung two large paintings of black panthers. It was a three-bedroom, and he moved his mother in with him, since he was the reason she had been forced out in the first place. Police raided the apartment every other month or so. From the lobby or the upper floors you could see them coming, and word would reach Cannon that the officers were on their way upstairs. He'd open his door and lie on the carpeted floor before they got there. No need to wreck anything or throw him to the ground. The cops might kick him in the side for being a smart-ass, but they knew they weren't going to find anything. And while he was arrested many times—for possession of drugs, intimidation, disorderly conduct—he managed to avoid another conviction. After the police raids, when Cannon was alone again in his apartment, he'd stare out the window, past the surrounding high-rises to the condos along Lake Shore Drive. "It looked like I was up there by the Gold Coast area," he'd say. "If you looked out my window, you'd swear you weren't in Cabrini. It was luxury."

8

This Is My Life

J. R. FLEMING

BEFORE CHANGING HIS name, in his early thirties, to Willie J. R. Fleming, he was Willie McIntosh Jr. People were always shouting, "Junior, Junior, Junior!" It made him feel small. Growing up, J. R. heard countless stories about his father. Willie McIntosh Sr. had volunteered for Vietnam and later went AWOL. They said he was Chicago's Frank Lucas, linking up with poppy producers in Southeast Asia and shipping the product stateside. He supposedly worked as a CIA operative in Nicaragua and who knew where else. They said he ran guns into Cabrini-Green. His nickname was Sweetness, and J. R. knew his father had to be slick to get his mother—a black-on-black-love, stop-the-violence war protestor—to marry him. "Let sleeping dogs lie," Marlene McIntosh would say when J. R. asked for specifics about his father.

J. R. was one of their six children, and when he was born, in 1973, they lived in the Henry Horner Homes. A few years later they came to Cabrini-Green. They lived in the 1150 N. Sedgwick building, the Rock, and J. R.'s father would sit out front with his friends, teaching J. R. how to fight, pitting him against other little boys as part of a training regimen, yelling at him to keep his guard up, to jab. Then, just as Jane Byrne was set to move into their high-rise, the McIntoshes left. Willie Sr. went to another of the red high-rises at Cabrini, with a girlfriend; Marlene took the children to the South

Side, first to the Robert Taylor Homes, where J. R. played on a softball team named Tuff Enuff, and then on to the cylindrical towers of the Hilliard Homes. Marlene had a city job, in Harold Washington's administration, and she was offered the chance to leave public housing and relocate with a rental voucher to the southern suburb of Dolton, a nearly all-white village about twelve blocks beyond the Chicago city limits. Marlene imagined she could give her children a better life there, and that's how J. R. ended up in a small house in the suburbs.

They lived on a tree-lined street of single-family bungalows. Their home had yards in front and back, a basement, an attic, and a porch. The meadows of a large public park began at their corner. "It was beautiful," J. R. said. They were one of three black families from 142nd Street all the way to 151st, and sometimes while they walked to school people let their dogs loose to chase them. J. R. was never one to lack for confidence, though. In Dolton, he earned money raking leaves and shoveling snow and delivering newspapers. Determined to prove he was at least as smart as the white kids, he read through the encyclopedia, A to Z, studied almanacs, and challenged his teachers. When a middle-school administrator told him he couldn't "court" white girls, he was undeterred. When someone spray-painted "Niggers Go Back" on their garage, he told the older boy next door, an acne-scarred white drag racer known as Big John, and they hunted down the perpetrators. Big John brought along a shotgun and blasted out the windows of the other boys' cars. J. R. grew big himself—six feet and two hundred pounds by the start of high school. He played basketball in the winter and tennis in the spring. But it was football he liked the most. The coach said they didn't do race on their team; they relied on one another, and all that mattered on the squad was kicking the asses of the guys on the other side of the line. It seemed to J. R. that the entire town came out for Friday night games. There were Alumni Club pancake breakfasts and Sundays huddled in the dens of his teammates to watch the Bears.

His sophomore year, J. R. was the starting running back. They called him Mac Attack, and he'd strut around in his letterman jacket. When he walked home after practice, still wearing his pads, people would call out greetings from passing cars, shouting that their nephews or neighbors' sons were on the team, too. The players needed to keep a C average or higher, but J. R. was a member of the academic honor society. He aced the college boards, a 29 on the ACT, thinking he was destined for college ball. And although he gave weed and coke each a try, he believed that he needed to keep his body pure for sports. He'd explode on the heavy metal boys in their Megadeth T-shirts smoking in the school bathroom. "Don't you know I'm a recovering asthmatic? Do you want me to have a bad game? Are you working for the other team?"

J. R.'s older sister, Joyce, moved back to Cabrini-Green. Their father was still there, along with their mother's sisters and cousins. On weekends, when there were no games and his mother was working, J. R. took the bus and train north to visit. His cousin Greg, known as G-Ball, ran with the Disciples in 1117 N. Cleveland, Dolores Wilson's old high-rise, which residents now called the Castle. J. R. would hear from him accounts of Castle Crew friends and enemies killed, of being shot at or shooting someone else. Back in Dolton, J. R. would lie awake nights thinking about Joyce, imagining the danger that engulfed her in the projects.

But there was a certain kind of aloneness to the suburbs that also filled J. R. with dread. He craved an audience and pressured himself to succeed in part to keep people around him. The better he was at sports, the larger the crowds that wanted to hang out. He learned to tell stories of his exploits, to entertain, megaphoning his words, a whirlwind of pantomimed emotions. In Dolton, neighbors would pull into their driveways after work and close themselves indoors. Adults socialized only now and then outside their fenced-in backyards. When the street baseball or the backyard basketball ended at dusk, the other kids went in for dinner and the block fell silent. For J. R., that wasn't a peaceful quiet. So the summer after his junior year, when he needed to rehab an injured knee, he asked to stay with Joyce at Cabrini-Green. He'd transferred to a high school not far from Dolton; the team played in a higher division, with more college scouts watching. The plan was for him to go to a physical therapist near Cabrini and spend a couple of months in the city as he readied himself for fall football.

For seventeen-year-old J. R. living without adult supervision, Cabrini-Green in the summertime no longer looked scary. It looked like paradise. It was 1990, and young people swarmed about, packing the swimming pools and baseball diamonds and basketball courts. Boys rode in front of the buildings on decked-out Schwinns, popping wheelies. Older guys detailed their souped-up cars. There was music—Janet Jackson, Digital Underground, Bell Biv DeVoe's "Poison." J. R. played basketball and softball. And at Cabrini, people didn't go inside at dusk. They partied outside the buildings, and not just on the weekend. There were so many girls. J. R. had dated in Dolton, but nothing like this. He'd ask a girl over to Joyce's to order up music videos on the television show *The Box*. They'd phone in their selections, picking rap videos for 99 cents a pop that weren't being shown on MTV or VH1, and chill for the night. When he didn't stay at Joyce's, he slept at an aunt's or a cousin's. Willie Sr. had moved to Alabama by then, but it still felt to J. R. that everyone was either family or like family.

J. R. got drunk for the first time. He smoked cigarettes. He got high and stopped going to rehab. One night near the end of August, he was coming back from a party in another high-rise, walking with his cousin, and he fell into the bushes by the basketball courts in Durso Park—Gangster Park, they called it there. He'd been guzzling Olde English 800, and he laughed at how drunk he was. When he came to hours later, his eyes level with the asphalt, he looked up to see the yellow-red glow of dawn. He realized then that he was totally free. "It's over," he bellowed to no one and to the world. "This is my life! I'm never going back to the suburbs."

■ ■ ■

IN 1987, HAROLD Washington won reelection, halting a Jane Byrne
comeback in the Democratic primary. Then he trounced his nemesis,
Ed Vrdolyak, who had switched parties to run in the general election. Gaining majority control of the city council as well, Washington
was positioned finally to push his agenda. But just months into his
second term, he slumped over his desk in city hall. At sixty-five,
he was dead of a heart attack. A new mayor, Eugene Sawyer, was
selected by his fellow aldermen after a long night of infighting. Hoping to resuscitate the moribund Chicago Housing Authority, Sawyer
turned for help to the Metropolitan Planning Council, the influential
nonprofit that had shaped city planning and the CHA from its incep-
tion; Elizabeth Wood had led the civic group before taking over the
housing authority. The MPC advised Sawyer to pick a new leader for
the CHA, the former Illinois governor Dick Ogilvie, who promptly
accepted the post and then also died of a heart attack. The MPC
offered up a second choice, Vince Lane, a forty-six-year-old African American developer of low-income housing who'd already been
chairing the organization's public housing committee. Lane took
over at the CHA in 1988, managing to be named both the executive
director and board chairman, positions that were ordinarily separated as a formal check on power.

A large, excitable man with a high forehead and trim mustache,
Lane had served as an executive of the Woodlawn Organization, de-
veloping nonprofit housing on the South Side, and he'd also started
a successful company building HUD-financed properties across the
country. But his conception of public housing had been formed earlier, in the 1950s, when he was growing up in a tenement by the
White Sox's ballpark. Living across the street from a low-rise public housing development called Wentworth Gardens, Lane envied
the working families there. Their homes were new, with neat gardens and large communal lawns. Then Lane watched in awe as the
twenty-eight towers of the Robert Taylor Homes rose up just east of

him. It was an architectural marvel and represented a marked improvement over the slum they replaced. One of Lane's aunts fled a tenement for the Taylor Homes, and the entire family celebrated her move as a major step up.

Lane was aghast at how far public housing had fallen. A man prone to pronouncements, his high-pitched twang would intensify along with his indignation. "Horrendous, horrendous," he screeched about the state of the CHA that was now his responsibility. The mismanaged agency was $30 million in debt, with a bloated staff and a stock of aging housing that hadn't been properly maintained in years. He heard from tenants who were so fearful of stray bullets that they slept in their bathtubs. In his first weeks on the job, gang members firebombed an apartment at Rockwell Gardens, a West Side development, wounding a little girl. At that moment Lane decided the hell with any real estate or management theories. He couldn't wait for permission from the mayor or a judge. It was his duty, at the very minimum, to establish public order and safety in the housing he oversaw. He borrowed an idea from a Korean War movie he liked called *Pork Chop Hill*. The GIs in the film, led by Gregory Peck, have to win back the bluff from the Chinese forces, or it's all over for them. As in the movie, Lane wanted to send in troops, overwhelming the enemy in each tower with a massive show of force. "I saw these high-rise buildings as my Pork Chop Hill, and I had to get control of them," he'd say.

At 5:00 a.m. on a Tuesday, CHA personnel and more than sixty police officers met at a staging area along the Dan Ryan Expressway. No one but Lane and the police chief knew about the target of the first raid, as they feared internal leaks would tip off the gangs. The convoy rolled west, to Rockwell Gardens. Outside the selected highrise, teams of police blocked the entrances and exits as dozens of other officers rushed inside, searching every apartment for weapons and drugs. CHA staff checked rolls to see who in the buildings officially belonged and who needed to leave. A second battalion of more than a hundred workmen arrived and got to work reinforcing the lobby with iron grating. They erected guard stations and installed video cameras. A midnight curfew was imposed, and residents were issued identification cards they now needed to display to reenter their building.

Lane called the action Operation Clean Sweep, and announced that he would conduct similar emergency "sweeps" at each of the city's 168 public housing high-rises and some of the low-rises as well. He would win the war in public housing by going door-to-door, overtaking one building and then another until the entire system was again safe. When Lane and his troops hit their second targeted high-rise, the surprise factor was already gone. "The lookouts saw this huge caravan of trucks and cars, and they realized we were coming to that building," Lane would recall with a laugh. "Drugs and guns rained out of the building." Within weeks, a parade of vehicles carrying 150 housing officials and police officers made its way to Cabrini-Green, to lock down one of the high-rises. Lane soon expanded these operations, adding tenant patrols and creating a separate CHA police and security force, its officers training alongside the Chicago police. He placed Chicago police substations inside each of the large housing developments, and had them pursue a kind of "broken windows" strategy, arresting residents for minor offenses as a way to prevent more serious ones.

Some CHA tenants and their advocates objected to the siege tactics. Lane never sought warrants to search people's apartments or to seize contraband, and when the American Civil Liberties Union filed a class-action lawsuit against the CHA, the agency had to agree to a legally binding consent decree setting limits on the raids—limits that Lane would exceed repeatedly. But other public housing residents welcomed the sweeps. They weighed the trade-off between personal freedom and police protection; they wanted the same level of safety as in every other Chicago community, the same standard of law and order. Lane would visit a development, and women would embrace and kiss him; men grasped his hand, thanking him for taking on the challenge. Residents submitted requests for the police to raid their buildings—or more commonly, the buildings next to their own. At Cabrini-Green, tenants reached out to Alderman Natarus, who then passed along the entreaties. "You and the police deserve thanks for leadership during the Cabrini-Green public safety crisis now taking place," he wrote to Lane. He asked that the agency put a stop to the cross fire between high-rises, adding, "You have my full support."

For many outside public housing as well, Lane's tactics seemed the appropriate response to the perceived mayhem of the inner city. The sweeps coincided with the hysteria over crack cocaine and with gangs battling one another over the growing market. The crack epidemic was real—by 1991, 70 percent of males arrested in Chicago tested positive for cocaine. Yet the crisis didn't elicit calls to mobilize medical and social services, and addicts weren't treated with sympathy. Rather, they were crack fiends, a scourge, and when they weren't being ignored in their ghettoes they were to be arrested, incarcerated, and eradicated. It was time for civil society to take its stand and reclaim its streets.

The teenage homicide rate rose steadily from the mid-eighties and into the start of the nineties, a tragic effect of the booming drug trade, the poor economy, and the growing strength of organized gangs. The number of killings in Chicago climbed each year, nearing numbers not seen in the city in two decades. But there was also much fretting over illusory threats. Images proliferated of premature "crack babies," infants exposed to the drug in utero and now convulsing in their cribs. The media reported constantly on this coming tide of damaged children spreading out from the country's black slums, a permanent underclass that would deplete the store of social services and grow up to run amok. In 1989, in New York City, a twenty-eight-year-old white investment banker was out on her nightly jog in Central Park when she was raped and beaten nearly to death. Police investigating the "Central Park Jogger" case arrested several fourteen-, fifteen-, and sixteen-year-olds who lived in Harlem. According to their confessions, they admitted to being out "wilding." The five teens never uttered the term; it was a police construct, made up during a night of bullying interrogations and deceptions, and the convictions would be vacated more than a decade later. But the term made sense to people, at least emotionally. It evoked a prevailing idea of out-of-control juvenile violence, the offenders like wolves in a pack, "savage," "feral," and "presocial." "How can our great society tolerate the continued brutalization of its citizens by crazed misfits?" Donald Trump wrote in a full-page ad he paid for in the New York *Daily News* before the boys' trials. "CIVIL LIBERTIES END WHEN AN ATTACK ON OUR SAFETY BEGINS!" John DiIulio, a political scientist at Princeton, coined the term "superpredator" to characterize this imagined growing population of murderous boys, mostly black, who were, he said, "fatherless, Godless and jobless." He cited a prosecutor who told him, "They kill or maim on impulse, without any intelligible motive."

In Chicago, the superpredators were coming for you from Cabrini-Green, obviously. "What is the place that quickly strikes fear in the hearts of the citizens of Chicago?" asked a 1988 profile of a police officer who dared to work the Cabrini-Green beat. "The Green has become synonymous with a 'jungle,' a place where wild animals roam and devour weaker organisms." A national television news program on Cabrini-Green asked viewers to visualize a barbarous war zone: "A no-man's land with broken windows, dark, abandoned buildings, no law and order. There are carefully demarcated areas controlled by rival bands of armed militia fighting over the rubble. Nearly every night there is sniper fire. It sounds like Beirut, but in fact it's America. A creature of state, local, and federal government. A product of bad politics, failed policy, and official neglect."

That the larger societal fears weren't based on sound science hardly mattered at the time. Children displayed no serious lasting effects from their prenatal exposure to crack; most babies born prematurely jittered in this way; and mothers who abused alcohol during pregnancy did as much harm and were far more common. Moreover, DiIulio's predictions that horrendous crimes committed by inner-city teens would double and triple in the coming years proved grossly inaccurate. Violent crime peaked in 1991 and then dropped precipitously; by the late nineties, juvenile crime rates had fallen back to the levels of the early 1980s, as the economy improved and crack use abated. By then, however, the hysteria had cemented into policy. Nearly every state had passed harsher sentencing laws for juvenile offenders; children were being tried as adults and receiving mandatory minimum sentences, including life terms. Pregnant cocaine users were charged with child abuse and even manslaughter. In Chicago, annual arrests for drugs other than marijuana tripled in less than a decade, with African Americans making up most of those charged.

It was during this panic that Vince Lane was touted as a hero. Republicans in Washington praised his aggressive approach, which was seen as a model for other cities to follow. Lane was compared to Joe Louis Clark, the bat-swinging principal of a New Jersey high school who was played by Morgan Freeman in *Lean on Me*, the movie based on his life. Clark had summarily dismissed hundreds of students who were, he said, "leeches, miscreants, and hoodlums." The country needed vigilantes like that who were willing to bend the rules. And Lane was tackling social problems that others thought too intractable and hopeless even to bother with, in the process giving CHA employees a new sense of purpose. Workers under him wore buttons announcing, "I'm Part of the Solution." He said the gangs who operated in public housing were like playground bullies, and the sweeps were his way to "throw the bullies out." In 1990, Lane was featured on the cover of the *Chicago Tribune Sunday Magazine* standing on the roof of a Cabrini-Green high-rise, other towers of tartan red brick rising up around him like battlements. Wearing a dark suit, his arms folded across his chest, Lane was a man reigning over his domicile.

But the sweeps were problematic from the start. In addition to being ruled unconstitutional, they were prohibitively expensive. The estimated cost of a single sweep of a single high-rise was $175,000, and that was without the amount paid in overtime to Chicago police officers. "It cost a shitload," Lane acknowledged. He was able to secure additional funds from Washington. The CHA also used money on the anti-crime initiative that could have gone to major repairs. The buildings suffered further decline. And whenever Lane swept a high-rise, crime increased in the surrounding buildings; he wasn't so much removing guns and drugs and miscreants as shifting them around. With each raid, fewer weapons and drugs were confiscated, as people anticipated the invading force. J. R. Fleming said news of the approaching CHA army would reach a building at Cabrini-Green and those with contraband would gather up their goods and store them outside in the trunks of cars.

Lane also knew enough about the world of public housing to understand that there was something counterproductive in removing every male who wasn't on a lease. Officially, women headed nine of ten households in CHA properties. Men were around, though. They were people's sons and lovers and fathers. "I was conflicted," Lane would admit. Who was he to say if the guys were drug dealers, or good fathers, or both? "To tear somebody out of this family environment was, I thought, wrong," Lane said. "It wasn't the right thing to do, the right signal to send. We wanted to send the signal that if you share a relationship with someone and have kids, we will work with you, help you get an education and all that. But I continued to move ahead and secure building after building. I really thought we could secure most of the city. Who would have thought that you could secure Robert Taylor and Cabrini-Green? Did I succeed?" He'd answer the question himself, his voice breaking into a higher register. "Absolutely. Absolutely."

J. R. FLEMING

BECAUSE HE WAS straight out of the suburbs, J. R. did run into some trouble at Cabrini-Green. Sporting his all-red football letterman jacket, with the collar popped up like Fonzie from *Happy Days*, he didn't realize that its color represented the Vice Lords and Cobra Stones. One time a Disciple near Jenner school demanded of him, "What d'you ride?" J. R. took the question literally, not even trying to be funny. "The bus?" He was at first perplexed that because he now stayed in a building controlled by the Gangster Disciples, he couldn't cross the blacktop and visit the building where he lived as a child that was now home to Cobra Stones. After the mostly white community of Dolton, he expected Cabrini-Green to be an all-black world free of tensions and bigotry. "What is this?" he asked. "Black people don't like black people?" When he attended Lincoln Park High School in the fall, everyone assumed he was in the Disciples because his cousins and next-door neighbors were in the gang. He would tell people that he wasn't Folks. "I'm a nerd. I'm an athlete. I'm competitive." It didn't matter. A group of Vice Lords chased him out of the pool, J. R. racing home in his swim trunks. Back on the land at Cabrini, people teased him, "You're not in Kansas anymore."

"Then tell me where I'm at, and what I need to do to survive," J. R. would demand.

He was walking one day, pushing a bicycle, when a couple of guys on Larrabee started telling him they were going to rob him. "Can I see your bike? I just want a ride, man." The bike belonged to his cousin G-Ball, from 1117 N. Cleveland, the Castle. No way was J. R. giving it up. He decided he needed to defend himself. This wasn't dunking a basketball or rushing for a first down. A family member lent him a .32 automatic. The following day he was heading up Larrabee with a girl named Tricia, and he told her about his predicament. "I got this right here for them," he said, showing her the pistol in his pocket. The weapon didn't surprise her. "Shoot them in the leg," she suggested as moderation. But when they came upon the guys, Tricia turned out to be related to them. "You're not shooting my cousins," she said, shouting, "He's got a gun!"

"You really got a gun for us?" laughed the bigger of the two, a guy people at Cabrini called Frank Nitti, after Al Capone's enforcer. He seemed amused, as if J. R. hadn't been in on the joke. "We were just fucking with you, man." He and J. R. would later become friends.

J. R. was inclined to see the world through the lens of sports, so he tried to approach Cabrini-Green as if it were a football game, discerning the rules amid the disorder. "You just need to follow your blockers. Follow your blockers and you'll survive," he told himself. He went to G-Ball and read through the Disciples literature. One of the gang's first tenets was that the GDs aided and assisted their own members. But J. R. looked around and saw that the gang was making a lot of its drug money in the Whites, the towers across Division Street; his red high-rise on Larrabee was on the front lines, directly across from a nineteen-story Vice Lords tower. It didn't seem like anyone was too interested in aiding and assisting them. J. R. phoned his father. Willie Sr. told J. R. to stay out of the gangs. He said J. R.'s brain worked differently than the brains of other guys in the projects, that he was too smart to get involved in petty battles. The advice played to J. R.'s ego. "Don't be a pawn in another man's game of chess," his father counseled.

Instead of gangbanging, J. R. formed a clique in his building called Skee Love, which was mostly about throwing parties, selling a little weed, and hooking up with girls. Their motto: "Bang, bang, bang. Skee, skee, skee." After being run out of Lincoln Park, J. R. tried another area high school, Wells, a mile and a half west. But he didn't see any learning happening there. Students brought guns and knives to school in their shoes. The Puerto Ricans fought the blacks who fought other blacks. One day a posse of Spanish Cobras was waiting for him outside the school. J. R. knew he couldn't fight the entire neighborhood. He elected instead to punch the assistant principal. He would be expelled. But he reasoned correctly that the police would come and drive him home, past the waiting gangbangers. It tore up his mother, but J. R. was done with school forever.

9

Faith Brought Us This Far

DOLORES WILSON

WHEN THE JOB of president of her building opened up suddenly, Dolores Wilson wrestled with whether to take it. Like the prayer said, she needed help distinguishing between what she could and couldn't change. Her own apartment she could control, but an entire building? The roof leaked and flooded apartments were left to rot. The top two floors were eventually closed off altogether. People weren't supposed to go up there, but they did. Gang signs were scrawled sloppily over the stairwells and elevators. Garbage collected on the landings. It got to the point where Dolores felt too embarrassed to invite anyone over. "I can't have visitors come through all of this to get to my house," she'd say.

For many years the president of the tenant council in 1230 N. Burling was a friend of hers named Ethelrene Ward. Ms. Ward worked with Marion Stamps to demand better services for tenants, and she helped start a community pantry. She let teenagers put on shows in the building, and she organized outings for the residents to the Wisconsin Dells and to a small lake where they played volleyball and fished, though Dolores couldn't stand to put the hook through the worm. When Ms. Ward died, a reverend who lived in the building took over. But he'd been discovered recently shot dead in an elevator. People said he'd gotten mixed up in a domestic dispute. Dolores believed that preachers should be judged not by their actions, but by their inspiring words, and she liked how at funerals the reverend would say, "I'm not talking to the deceased. She can't hear me. I'm talking to all of you in the audience." But now he was the deceased, and she was still seated in the pews. She decided to take on the responsibility and succeed him as president.

She called her first meeting about a year after Hubert's death. Debbie and Cheryl, her daughters, both had their own apartments in the building now, so Dolores knew they'd show up. But she figured few others would come. When she walked into the first-floor rec room, though, she discovered it crowded with seniors and single mothers and little children and even some gang members. The older tenants talked about restoring 1230 N. Burling to its original glory in the early sixties, when it was freshly painted and surrounded by green grass, flowers, and benches. The young parents complained about the gangs charging them to ride the elevators, about their children being beaten up and robbed. And the gangbangers who were present agreed for the most part—they, too, were offended by the conditions in the high-rise. One of the gang leaders raised his hand and suggested putting stools on each floor, since children who couldn't reach the trash chutes were spilling garbage or leaving bags on the ground.

A dozen people at the meeting volunteered to monitor the building in shifts. All but one of them was a woman, and many of their neighbors called them crazy, saying they were going to get shot. But they went up to the gang members and told them to set up someplace else. They worked in teams, one stationed by the back door leading to Halsted Street, another watching the front, and others on the elevators. When Kelvin Cannon moved into the Burling building after his release from Stateville, he took a shift as well, safeguarding the elevators even as he led the Disciples. Dolores pressured the CHA into replacing lights in the lobby and installing a phone there. One woman sold candy out of her apartment, and with the proceeds the security team bought blue uniforms with a 1230 BURLING patch sewn onto the sleeve. Eventually the CHA paid some of them as security guards. They carried walkie-talkies and clipboards, keeping a list of every unlocked, vacant unit and broken window to send on to the housing authority. "This is where we live, and we have to protect it," Dolores said. "We started laying down the law, and we found out the law was on our side."

The CHA allotted the building a small monthly stipend for miscellaneous expenses, enabling Dolores to outfit the rec room with a used pool table, a Ping-Pong table, and a thrift shop sofa with a psychedelic design. She convinced a local store to donate ten gallons of paint, and she paid men from the building $6 for each stairwell they painted. Some of them were so eager for work that they did two or three floors on their own. In the summer, the building hosted a rummage sale, and residents hung clothes on the fence outside; they set up grills and sold hot dogs and refreshments. A couple of tenants planted a garden on the empty land alongside the building. There was a fashion show, a coat giveaway, and a tutoring program.

Dolores conscripted children to help with the tower's upkeep. The same boys and girls who might have fooled around on the elevators now wore pieces of paper with the word MONITOR taped to their shirts. They walked the high-rise in pairs, making sure other children didn't litter, prop open the elevator doors, or jump out of the cabins. Mayor Washington's administration gave the building a permit to block off Scott Street, just to the north of the high-rise, and the residents held a jump rope contest and a Big Wheel race. Because she hated competitions, the whole idea of grown people yelling when a ball didn't go in a net or a glove, Dolores awarded each child a prize. Neighbors told Dolores that they never imagined they'd clean up after someone else, but the building was looking so good that they had reached down and picked up a potato chip bag from the lobby floor. Dolores soon felt comfortable enough about the state of the high-rise to invite her pastor from Holy Family over for tea. And it thrilled her when he said from her living room that 1230 N. Burling looked like no other building in Cabrini-Green.

Dolores still had her full-time job at the city water department as well as her other volunteer work on local boards and committees. As the secretary of the tenant council for all of Cabrini-Green, she was taking the minutes at a meeting in 1985 when the talk turned to a curious pilot program being funded by the Metropolitan Planning Council. The civic nonprofit wanted to train tenants in a few public housing developments across the city to take over management duties from the CHA.

In recent years, the MPC had considered recommending that all the high-rise public housing in Chicago be torn down. But during the 1980s, both poverty and the number of families in need of low-income housing increased in Chicago while federal dollars declined and affordable housing options disappeared. The homeless population in Chicago and the number of people on the CHA waiting list swelled. "Given those circumstances, Chicago cannot afford to lose any housing for low-income families, including the 38,685 units of public housing," the MPC concluded. Instead, the organization explored the resident-management model. The premise was that tenants would be more motivated than housing authorities or private managers to renovate vacant apartments in their buildings, collect rents, and deal with bad neighbors. And besides, nothing else had worked. The resolution that the Chicago City Council passed to create what it called a "vision of a brighter possible tomorrow" read more like a public housing postmortem:

WHEREAS, the abysmal quality of services rendered to residents of Chicago Housing Authority by management is the stuff that horror stories are made of—tracts of fifteen- to nineteen-story buildings where at times elevators have been inoperative seventy percent of the time—residents waiting for years for new shades, for screens for their doors and windows, waiting months for replacement windows and the most simple repairs; and . . .

WHEREAS, the wholesale flight of working families from
Chicago Housing Authority developments . . . have left it a
vast urban wilderness three-quarters of whose adults are
unemployed, eighty percent of whose households are headed
by unemployed single mothers . . . where welfare dependence
and the welfare mentality hangs like inescapable miasma in
the air coexisting with the all too frequent incidence of criminality, marginality, gang violence, alcoholism, drug abuse
and other self-destructive behavior bred by the unbroken
landscape of poverty and desperation . . .

The idea that residents of public housing be put in charge of
their own buildings was born not only out of desperation but also
from an unlikely union between conservatives and black activists.
Resident management meant jobs and skills for African American
tenants that might lead to success outside public housing. This self-
empowerment aligned with the Right's ideal of personal responsibility and a diminished government, what Republicans decried as the
trillion-dollar "poverty-industrial complex" established under Lyndon Johnson's War on Poverty. Chicago wouldn't be the first city to try
the experiment. In Boston, St. Louis, Jersey City, and Newark, public housing residents had already formed companies and taken over
maintenance and daily operations in their own buildings, although
the results had been mixed. Housing authorities remained wary:
landlords and building managers were supposed to be allies, partners in operating any property, yet in public housing, agencies had
long had an adversarial relationship with their tenants. Residents,
too, had little to no accounting, hiring, or supervising experience,
and the buildings they were put in charge of managing were among
the worst in their cities. But tenants embraced the opportunity, and
conservative think tanks pushed to expand the venture to other cities. When the Metropolitan Planning Council surveyed residents
of Cabrini-Green and two other public housing developments in

Chicago, it found that many of the unemployed single mothers would much rather work than receive benefits; they'd been trained as seamstresses, cosmetologists, cashiers, teachers, secretaries, and cooks. Here was an untapped pool of labor. "The brave pioneers of resident management are sowing the seeds of hope and possibility in cities across our nation," Ronald Reagan declared from the White House.

Dolores Wilson didn't think too much of the idea when she first heard about it. "I didn't even know what resident management meant," she'd say. But then someone from the MPC phoned her at the water department. Because the people in her high-rise had shown initiative, because they'd already done so much to better their home, the person hoped that Dolores and the other residents in her building would apply for the program. Dolores sought the advice of one of her neighbors, Cora Moore. A mother of six, Moore worked for the CHA as the assistant manager of 1230 N. Burling, and had lived at Cabrini-Green since 1969. "Yeah, yeah, we want resident management," she shouted, saying this was the opportunity they'd been waiting for. Dolores signed them up.

THERE WAS A ten-year-old from the Reds named Rodnell Dennis. He didn't have J. R.'s way with words or athletic abilities; he wasn't shrewd or confident like Kelvin Cannon. But he was big and powerful for his age, and he'd acquired the nickname Dirty Rod. He'd "do dirt" on other little kids, grabbing them by the neck as they were heading to the candy store on Orleans Street and robbing them of their change. Also, he'd admit himself, he was plain dirty. His mother got hooked on drugs and stopped mothering. His father wasn't around. The family had little money, and his clothes were raggedy and unwashed. Rod was surprised one day in 1989 when a Gangster Disciples coordinator told him the gang had been checking him out for a few years and liked what it was seeing. He said the GDs didn't usually bring young fellas into the fold, but they recognized in Rodnell a fearlessness and believed with a little guidance he could be someone. A boy starving for sustenance of all sorts, Rodnell was spellbound by the praise. He was introduced to the GDs operating on the Wild End, the southern portion of Cabrini-Green that included some of the red towers and the rowhouses. "What do you feel you can do for this organization?" a GD governor asked him. "I can do whatever you want me to do," Rodnell shouted too quickly, and the group of young men around him laughed.

The gang members saw that Rodnell lacked for many things, and they gave him new shoes and clothes, all in the GD colors of black and blue. In return, he was expected to make certain sacrifices. For someone at the bottom rung of the organization, that meant security time. Every day after school, from three thirty until nine, he stood outside one of two high-rises. Others sold drugs from inside the building. His job was to yell, "One time, lights out," when the police approached. The Disciples always talked about the Five Ps— "proper preparation prevents poor performance." Rod now lived by this credo. He attended regular meetings the gang held at the Lower North Center, in the gym or pool room, where they'd discuss business and neighborhood events of the past week. They were instructed that it was their job to take out their enemies. They were at war with the Vice Lords and the Stones in the nearby towers. A higher-up would say, "Who wants to go on a 'mission'?" Rodnell, eager to show how motivated he was, would be among the first to volunteer.

When he was thirteen, in March 1992, Rodnell went on a mission. It was a freezing Sunday afternoon, and one of Rodnell's best friends, a boy who would soon be killed, stepped into a circle of GDs. He held out a handgun, a silver .22. He'd spotted a group of Vice Lords not far from Jenner school. Who among them was going to teach the Vice Lords a lesson? Rodnell reached for the gun. He ran over to the 500 W. Oak building, a couple hundred feet away, and ducked behind a Dumpster by the old Death Corner. People were gathered outside the front entrance of the high-rise. Rodnell could hear the sound his gun made, feel the recoil, but otherwise his mind went blank. When he realized he'd emptied the chamber, he ran back to his building, handing off the gun to another boy, who hid it inside an elevator shaft. He took a seat on the stairs outside the tower, breathing heavily into the cold air, and tried to act as if nothing had happened. And maybe nothing had. Less than ten minutes had passed since he'd grabbed the gun, and here he was back where he'd started.

For most shootings, a cop car or maybe two showed up. But ten now raced down Oak, along with an ambulance. Rodnell had to check out the commotion, and he returned to the scene of the crime, sauntering over as casually as he could. Yellow tape surrounded the building's entrance and a white sheet was draped over what had to be a body. The thought of killing someone hadn't actually entered Rod's mind. The victim, he learned, was a boy named Anthony Felton, a nine-year-old who had no part in the Vice Lords or any other gang. Rodnell knew the child's family. He retreated to his building. Two police officers were already up on his ramp talking to his mother. Someone must have identified him as the shooter. He saw a police cruiser and ducked into one of the high-rises.

There were two plainclothes cops at Cabrini-Green whom everyone called Eddie Murphy and 21. They were young and black, and they treated the residents with respect, even the guys they arrested, referring to them as their "clients." When things were calm at Cabrini, the officers would visit with families in their apartments. They lifted weights with guys, helped run a baseball Little League, and challenged boys on the basketball courts to games of two-on-two. People at Cabrini believed that Eddie Murphy and 21, unlike other cops, wouldn't slap them around or plant dope on them or pocket their money. And when a guy committed a serious crime and had exhausted his options, he sometimes refused to turn himself in to anyone else but them. Now the officer known as Eddie Murphy found Rodnell's cousin and asked him to pass along a message: "If Rod runs, we'll have to shoot him."

Rodnell didn't have anywhere to run. He was thirteen and all he knew was Cabrini-Green. The next night, sleepless and panicked, he walked over to the police station on the ground floor of the 365 W. Oak high-rise and banged on the cage. An officer asked him what he wanted. "I heard you all were looking for me."

"Who are you?" the cop asked.

"I'm Rodnell Dennis. I'm the one that shot that kid." Because of the record he'd amassed since the age of seven—criminal damage to property, petty vandalism, theft, attempted theft, shoplifting, battery, robbery, possession of a weapon, auto theft—he was transferred from juvenile to criminal court and tried as an adult. He pleaded guilty and was sentenced to thirty-nine years.

EDDIE MURPHY AND 21 were James Martin and Eric Davis. Martin was working as a beat cop on the South Side when he issued a couple of traffic tickets to Jesse Jackson Jr., the future disgraced US congressman and son of the famous reverend. After ticketing Jackson, he was transferred to Public Housing North, the commander at Cabrini-Green greeting Martin with bemused laughter. On that very first day, he was in a squad car on Oak Street when rival gangs in the red high-rises opened fire on one another. A man jogged over from the rowhouses to join the gunfight, using the police cruiser with the cops inside it as cover. "I'd never seen anything so ignorant in my entire life," Martin said. He'd grown up in public housing himself, at the South Side's Ida B. Wells Homes. But he'd had a peach tree outside the rowhome he lived in with his grandparents; he'd graduated high school and attended West Point for a year. Tenants nicknamed him "Eddie Murphy" because he had a passing resemblance to the comedian and also because he cracked jokes endlessly, even making fun of a guy's clothes or running style as he was clasping cuffs on him.

Eric Davis was so fresh faced when he showed up for work at Cabrini-Green in 1987 that residents called him "21," for the TV show *21 Jump Street*, about cops young enough to go undercover as high school students. He'd lived in a Cabrini high-rise as a child

in the 1960s, when his family first made the trek north from South
Carolina. He went on to become a prep basketball and football star
in the Uptown neighborhood, and then as a backup point guard he
co-captained the University of Houston basketball team—Phi Slama
Jama—that featured Hakeem Olajuwon and Clyde Drexler and lost in
the 1982 Final Four to Michael Jordan's University of North Carolina
Tar Heels. As a cop, Davis wanted to serve at his former home.

Crack cocaine was just starting to ravage the neighborhood.
Mothers, long the cornerstone of the community, were disappearing
for days, handing off their children for others to watch. In Cabrini-
Green's white high-rises, a renegade faction of Gangster Disciples
bought drugs from suppliers not sanctioned by the gang's leadership
and kept the profits. Across the dividing line of Division Street, you
didn't just have Disciples fighting Cobra Stones and Vice Lords but
also GD hit squads firing at one another with automatic assault rifles.
And the Disciples in the Whites warred as well with young men in a
rival gang who lived a couple of blocks away in the Evergreen Terrace
apartments, a small low-income development next to Marshall Field
Garden Apartments on Sedgwick Street.

As a way to talk to the kids at Cabrini-Green, and as a response
to the gangsta rap that was then popular, Davis, Martin, and another
officer formed a rap group of their own, calling it the Slick Boys,
slang for undercover cops. A sometime drug dealer from Cabrini,
Pete Keller, known as K-So, helped them with lyrics, which were part
rap battle rejoinder and part public-service announcement. "C is for
CHA, which really lacks / D is for the dealers of drugs like crack
/ E is for the end of the ee-conomee / F is for the fathers that I'd
rather see." The media embraced the story of the police trying to save
lives through music, with headlines such as "Cops Stay on the Beat."
The Slick Boys delivered talks at public schools about the dangers
of gangs and drugs. They worked nights at Cabrini and then by day
traveled around the city and later the country to deliver their musical
message about the efforts needed to turn around blighted commu-

nities. They hired dancers and roadies from public housing. They filmed a music video. A movie was written based on their lives, hinging on an invented beef between a fictional drug lord at Cabrini-Green and 21, the two presented as childhood friends now on opposite sides of the law—and the mic!

In the summer of 1991, Eddie Murphy and 21 learned that a security guard in one of the Cabrini high-rises had kidnapped a thirteen-year-old girl from the building. Michael Keith was twenty-six and had worked for the private security firm for half a year. The girl was Veronica McIntosh, J. R. Fleming's baby sister, who along with their mother had moved back to Cabrini-Green that June. At 1017 N. Larrabee, where they now stayed, Veronica jumped rope with other girls late into the night. One afternoon, she and her fourteen-year-old cousin were in front of another Cabrini high-rise when Keith pulled up alongside the girls and offered them a ride. They knew him from the building, so after hesitating for a moment they climbed in.

Keith drove a white Oldsmobile Cutlass, a two-door coupe, and Veronica shimmied into the back while her cousin sat up front. For a while Keith was chatty and amiable, pointing out stores and buildings as he drove them around the Near North Side. It was when the girls said they were ready to go home that he turned strange. Between long silences, he told them that wasn't what they wanted. At a stoplight, Veronica's cousin jumped out. But as Veronica was sliding out as well, Keith grabbed her and pulled the door shut. She turned to face him. He held a gun pointed at her face. Over the next twenty hours, with Veronica crouched against the red leather backseat, Keith drove silently. He'd park somewhere secluded, climb into the back, rape her, and drive some more.

Veronica hardly knew the city, so she had no sense of where they were. During the night, Keith stopped outside a house in a residential neighborhood and tried to get her to come inside. But she clutched the seat and screamed, and he gave up. Later Veronica learned that it was Keith's home on the South Side. She'd blame herself for not going inside with him. J. R. had been with a girl when Veronica was abducted, but he'd since tracked down Keith's home address through the security firm. If she had just gone inside, Veronica told herself in the days and decades to come, then J. R. might have saved her from a dozen more hours of hell. It wasn't until the next day, with Keith bleary and muttering, that Veronica slipped out of his reach and fled the moving car. She ran away along a crowded street. She looked up and saw the Cabrini high-rises in the distance and headed toward them. Keith didn't return to his house, but he did come to work to pick up his paycheck, and Eddie Murphy and 21 were waiting for him.

Years later, after Keith had served a fourteen-year prison sentence, Veronica confronted him in a South Side diner. Up to then, she avoided talking about what she referred to only as "the incident." She'd lost her ability to trust people. She believed all men were devious, and she had trouble leaving her own children alone even with their father. For a while she'd popped pills and turned violent. With her sister by her side, Veronica asked Keith how he could have done what he did to her. She was thirteen, a baby. She'd still had dolls lined up in her bedroom. He'd stolen her childhood. Keith didn't apologize or ask for forgiveness. Rather, he smiled coyly as he reflected on their hours together. He said he'd really wanted her. And since it was public housing, he took her.

DOLORES WILSON

For DOLORES WILSON and the other public housing residents training for self-management, it felt like they had enrolled in school. "What we did took longer than college," Dolores would say. They spent hours in workshops and evening sessions, studying best practices for screening tenants and for conserving heat and electricity. They learned how to read a lease. They reviewed the endless rules and regulations issued by the CHA and HUD. They went over how to form committees, bill vendors, and fill out tax returns. "I think

I came to know more about housing than Jack Kemp did," Dolores said, referring to the NFL quarterback turned Republican senator turned secretary of housing. They went on weekend retreats outside Chicago, where a trainer taught them leadership skills. As managers, they couldn't lord their power over others. They needed to understand that it was better to listen than to speak all the time. In their new positions, they would have to inspire their neighbors. "If no one is following, you aren't a leader," the teacher repeated. But it was also going to be up to them to lay down the law, to put an end to illegal activities in their buildings. They role-played how to resolve conflicts and also to evict tenants whose conflicts couldn't be resolved.

The Metropolitan Planning Council called this effort "empowerment training," teaching residents how to become "participants in the process of managing their buildings and deciding their future." To model this self-determination, the MPC hired Bertha Gilkey, a resident of a public housing complex in St. Louis. In 1969, Gilkey, twenty and a single mother, organized a system-wide rent strike that lasted nine months, and she went on to run the tenant group at Cochran Gardens that took over custodial and management duties from the St. Louis Housing Authority. Cochran Gardens had been a lot like Cabrini-Green: abutting a gentrifying city center, it was seen as a "war zone," with "rooftop snipers and drug wars." But under resident management, the rehabbed apartments filled to near capacity, crime fell, and rent collection doubled.

Dolores liked Gilkey instantly. The sessions the St. Louis activist led were as much self-help revival meetings as they were classes in operations and finance. Gilkey's voice was husky and operatic; her hair was cut into an asymmetrical bob, her eyebrows arched in a look of perpetual defiance. She would boom that the tenants had been mistreated and misunderstood for too long, dismissed as pushers, pimps, and dope fiends, as if the poor were without dreams or aspirations. She made the Chicagoans believe in themselves and their desire—their *right*—to run their own buildings. She'd chosen against being on welfare, she explained to her trainees, a personal act of willful defiance. And she expected as much from them. "What I'm saying is that there is a new day, that you will no longer be crying in the wind," she pronounced. "1230 Burling is going to be a decent, safe, clean, and sanitary place to live." Gilkey would bring them into a tight circle, all of them clasping hands, and lead them in song. *Mine eyes have seen the glory of the coming of the Lord!* "She was revolutionary," Dolores would say. "When Bertha spoke, she made you listen. You had to."

Gilkey emphasized that outsiders wanted them to fail, expected it. Others had been enriching themselves on their backs and wouldn't give that up without a fight. "There's big money in poor folks," she preached. She challenged them to look over the bloated staffs at the Chicago Housing Authority, the deputy deputies and the executive executives. She said that's where the finances meant to improve their buildings had been going for decades. Housing authorities were paid by HUD for units whether they were vacant or filled, so the agencies had no incentive to make their developments better. The CHA had failed to carry out its charge of providing safe harbor for those overlooked and exploited by the private real estate market. It was up to them, the tenants, to change things. Otherwise, they would continue to live in degradation, or worse, on the streets, because cities were going to generate the will eventually to demolish their run-down projects. She'd seen it happen in St. Louis with Pruitt-Igoe. But she was going to show them how to succeed. "The power to control your own destiny is freedom," she proclaimed. Then they'd join together in shouting, "I'm fired up! I'm tired, and I can't take it anymore!"

Dolores and her neighbors were fired up, but they were also overwhelmed. The task ahead seemed both too tangled up in endless details and too abstract. When they felt about ready to quit, Gilkey took them on a trip to St. Louis. Fifty Chicagoans got off the bus at Cochran Gardens and couldn't believe that what they were seeing was public housing. Under a tenant-led renovation, Cochran Gardens had added new townhomes to the tracts of concrete that had separated its towers; the high-rise units were rehabbed, the buildings updated and outfitted with balconies overlooking a communal courtyard. Everything was pristine, even the incinerator room. Dolores, wearing her Harold Washington button, declared it unbelievable. Cora Moore nodded in approval, envisioning the possibilities at 1230 N. Burling.

In the spring of 1988, after eighteen months of classes, the trainees from 1230 N. Burling were honored at a graduation ceremony held at a downtown Chicago bank. Many of the residents put on their Sunday best—pink and yellow dresses and towering hats ringed by brims as wide as umbrellas. Dolores covered her mouth as she laughed and wept. "I can't succeed without you and more people like you," Vince Lane told the graduates.

Dolores wrote a letter to the newspapers, reporting on their achievement. "We have learned many things about tenant management," she noted. "Now we must put it to work and open the minds of *all* residents if we are to succeed." In 1990, the leaders of 1230 N. Burling were designated interim managers of their building. Then in 1992, after seven years of preparation, the 1230 North Burling Resident Management Corporation took over an annual budget of $6 million and became responsible not only for providing security but also for collecting rents, screening tenants, and maintaining the property. In a written statement, its members declared it their mission to "provide management programs and services, social, educational, cultural, and spiritual, to better the lives and living conditions of the 1230 North Burling residents." Dolores designed their personalized letterhead—a rough ink drawing of the file cabinet high-rise, the windows colored in to highlight the white facade, the motto "Faith Brought Us This Far" stretched across the roof and down one side of the tower.

Dolores served as the group's president, an unpaid position, and Cora Moore ran the day-to-day operations as the head manager. They created a seven-member elected board of directors, and hired a full-time paid staff of seven, which included a leasing clerk, an accountant, and janitors. They recruited two residents from each floor to serve as floor captains, and numerous other tenants joined the building's fifteen different committees. Wilson required members of the management team to look respectable, since they were now representatives of the building. "I'm not talking about going to the mailboxes in high heels and makeup," she said. "You just have to be decent. Don't come to the lobby looking like hell's a popping." Potted plants were placed by the mailboxes, and visitors now needed to be buzzed in to enter. The managers inspected every apartment and began the eviction process for tenants who didn't follow the rules. Dolores said they had to keep out "undesirables." They started a nursery in the building, covering cinder block walls with yellow and blue hearts. They opened a Laundromat on the second floor, negotiating a sixty-forty profit split with the company leasing the washers and dryers, the tenants selling tokens out of the management office, so no money accumulated in the machines. ("Cabrini Tenants Awash in Still Another Success," ran a headline.) They operated social service programs for young people and senior citizens. They partnered with a nonprofit to build a new $60,000 playground for the hundreds of children who lived in the tower.

People from other Cabrini high-rises started approaching Dolores, asking if she could get them an apartment in the building. But she didn't want anyone saying she played favorites, and she referred them to the admissions committee. The building was described in the press as "a ray of hope," "a shining example of grassroots empowerment." For Dolores, the highest praise came from Washington, DC. "President Bush named our building a model for the nation," she announced.

10

How Horror Works

J. R. FLEMING

AFTER J. R. Fleming was expelled from his high school, in 1990, for punching the assistant principal, his mother packed his suitcase and put him on a Greyhound bus headed for Alabama. No way he was going to hang around Cabrini-Green all day, not when they were talking about a murder epidemic in Chicago. J. R. had started spending time with a girl named Donna who lived on the third floor of his high-rise. He was seventeen, and she was fifteen and pregnant with their child. But J. R.'s mother still sent him away to his father. Willie Sr. was living in Alexander City, a textile town in Tallapoosa County, about halfway between Birmingham and Auburn. When J. R. stepped off the bus, his dad was there to greet him, standing alongside the white sheriff. The two men drove J. R. directly to the police station and showed him the jail. "I hope you're not trouble," the sheriff drawled. He wouldn't be, J. R. promised.

J. R.'s sisters had always been "dad hogs" whenever Willie Sr. visited Chicago, dominating his every waking moment. Now J. R. had the man to himself. He was put to work doing carpentry and landscaping. But he also passed the days riding around the county with his father, along the river and lake and past the many sewing factories for Russell sporting goods, which hadn't yet moved its manufacturing first to Mexico and then to Honduras. They shot pool together and shot guns at a range and drank in bars. His father talked about his experiences in and out of the military, both legal and extralegal. J. R. listened as his father swapped tales with a group of other Vietnam veterans, the men somehow seeming to run what they called Alex City. J. R. turned eighteen, and his father made him sign a selective service card, but he also told his son that government work probably wasn't for him. "Be your own man," he chided. And J. R. tattooed on his bicep a scratchy M.O.M. It looked like "mom" but stood for "My Own Man."

When J. R. returned to Cabrini-Green after five months, his life there seemed different, but also much the same. He partied with his Skee Love crew in 1017 N. Larrabee, but he had a baby now with Donna, a son they named Jonathan. He played in basketball games and in a softball league, leaping to catch fly balls in front of teammates who were hunched in standing slumbers from heroin highs. But the dreams of college sports that had defined his life up to then were over. What J. R. decided he needed to do was make some money. A guy named Joe Peery, who worked at a local youth organization, was impressed by J. R.'s brashness and intelligence. He helped J. R. land his first paycheck job, sorting packages by zip codes for UPS.

J. R. lasted less than three months at the job. The minimum wage was then $4.25. He couldn't buy Pampers or feed his son on $4.25 an hour. And he sure couldn't afford all the other things he wanted for himself. J. R. was obsessed with cars, and he'd always been drawn to technology, trying to test whatever gadget was new to the market, whether it was a computer, a PDA, a pager, or a video camera. But the taxes were what sealed it for him. When he got that first UPS paycheck and saw that $11 had been withheld, he was furious. He could make more money selling weed in three days than he did in an entire week sorting mail.

He decided to deal crack. The guys around him who had the cars and the game systems, who were eating like kings, that's what they did. J. R. asked around, but no one at Cabrini would put him on count, giving him the product to sell. Even from Alabama, Willie Sr.

still had clout at Cabrini, and he'd asked the local heads to keep his son out of it. G-Ball, J. R.'s cousin from the Castle Crew, told him, "This is what I do, so you don't have to do it."

J. R. was anything if not persistent. He convinced one of his Skee Love guys to give him some work, so long as they split the profits and word never got back to anyone at Cabrini. The friend had gotten a secret stash from a liquor store owner on Larrabee who said he'd gotten the drugs from a cop who worked on the West Side. J. R. headed way up north, far from Cabrini, walking a block that looked to him like a good place to sell the drugs without anyone from home finding out. He'd worked less than a day when the police nabbed him. They'd been staking out the street, building a case against the Cobra Stones there, and then this dumb, loudmouthed kid showed up. If they hadn't arrested him, they told J. R., he'd likely have been killed that night. The cops didn't include J. R. in their larger criminal indictment against the Cobra Stones, and as a first-time offender he avoided jail.

That was the end to his brief career as a crack dealer. In little time, J. R. devised a new moneymaking scheme. It was the spring of 1991, and people in Chicago were fiending as well for Michael Jordan and the Bulls. The team had just swept the Pistons in the playoffs, after losing to Detroit each of the past three years, and the Bulls were on their way to their first-ever title. Jordan, too, was ascendant, appearing in ads for McDonald's, Coca-Cola, Gatorade, Wheaties, Ball Park Franks, Edge shaving cream, Hanes underwear, Nike, and Chevrolet. The basketball movie *Heaven Is a Playground*, filmed on location at Cabrini-Green, was about to hit theaters, and Jordan had initially signed on to play the lead, pulling out after he'd become too famous. When the Bulls beat the Lakers for the championship that June, many of the stores along Larrabee were looted, people in their revelry and despair firing off guns, smashing windows, and taking whatever goods they could carry. Several of the local shop owners were Jordanian—from the Middle East country of Jordan, that is— and by then fixtures in the neighborhood. But when one of them tried

to protect his business, brandishing a gun in front of his store, someone in the melee took the weapon and beat him with it.

What J. R. did was use his unemployment money from the UPS
job, plus a little weed money, and buy a stack of bootleg Bulls championship T-shirts from a vendor on Roosevelt Road. Back home and
around the Gold Coast, he resold the shirts for twice what he paid, increasing his stock with each re-up. A man unpossessed of an inside
voice, incapable of speaking softly, J. R. was a natural salesman:
"Three for twenty-five! Got to sell today!" He'd bully, boast, sweet
wheedle. He didn't take no for an answer, wouldn't let customers turn
around. He'd do anything, he said, to get the dollar from their pocket.
Like any great salesman, what J. R. really peddled at all times was
himself. And in that product he believed spectacularly.

J. R. resembled the Bulls' backup forward Cliff Levingston, with
the same deep-set beaded eyes and wide-mouthed smile, the same
broad-shouldered build and leonine head topped with a hi-top fade.
At clubs around the North Side, J. R. pretended that he was Levingston's son, and he'd get in without a cover charge or score free
drinks. When the Bulls were in the playoffs the following season
and still weeks away from their second title, J. R. went all in on the
"Repeat" merchandise—the hats and shirts with the paired rings
or trophies. To increase his profits, he paid off the deliveryman who
brought the goods to the vendor on Roosevelt Road, giving him $700
for the name and address of the New York distributor. From there, he
connected directly with the wholesaler in Malaysia. He found another supplier who sold the hologram stickers that signified an NBA-
licensed product, affixing them to the knockoff gear. He hired guys
from his Skee Love crew, each of them pushing a shopping cart that
had been converted into a rolling sporting goods store, with wood
beams set up on three sides to display the wares. They sold the Bulls
merchandise by the Rock N Roll McDonald's, on Michigan Avenue's
Magnificent Mile, and outside the Near North Side's actual sporting
goods superstore. J. R. had recently joined the Young Democrats of

Cook County, showing up afternoons to clean out the ward offices or haul mats for the Jesse White Tumblers. When police officers told him to move along from wherever he'd set up shop, he'd recite the peddling laws he'd memorized. When that didn't work, he let the police know that they could contact Jesse White, Alderman Natarus, or even the mighty boss of the ward himself, George Dunne.

ANNIE RICKS

SETTLED INTO HER fifth-floor apartment in 660 W. Division with her kids and her mother, Annie Ricks refused at first to send her children to the schools in her new neighborhood. Every morning, she put them on a bus and train back to the West Side, an hour-long journey each way. "I knew the area," she would say of the West Side, so it seemed safer, more manageable. "I didn't know anything about Cabrini." But at Schiller, the principal welcomed Annie, assuring her that the Ricks children were going to love it there. He was right. They did love it. Ricks was soon volunteering in their classrooms. From the time she was a little girl in Alabama, she'd wanted to be a teacher. One of her cousins, already in Chicago, had a job in a public school, and that's what Ricks hoped to do, too. And when she was hired as a teacher's aide at one of the Cabrini-Green neighborhood schools, it was a dream come true, of sorts.

That was just one of her jobs. Annie also managed an after-school program through a local church. At three o'clock, she'd leave the elementary school, pick up art supplies and softballs, and lead two dozen children from her building to the blacktop. She'd shepherd the kids on long walks, as far as the beaches and Navy Pier, warning them that if they acted up she would never take them anywhere again. She babysat neighbors' children and worked at another local organization as well, running sports programs and helping Al Carter with his Cabrini-Green Olympics. Over the summers, she served free lunches out of the park by her high-rise.

Ricks took her own children across Cabrini-Green for further tutoring at the old Montgomery Ward headquarters. They went to after-school programs in the neighborhood churches and youth centers. And Ricks rarely missed one of their games. The Ricks were a basketball family, the girls as well as the boys starring on their teams. With thirteen of them, Ricks children filled out entire squads in pickup games at Seward Park. Her older children met Scottie Pippen when the Bulls star paid to renovate the basketball courts at Durso Park. Ricks gave birth to Rose during the Bulls' third NBA finals series, in 1993, and to Raymond during one of the subsequent repeats.

Annie's days at Cabrini-Green were long. At home, Ricks might fool around and wrestle with her children. She loved the professional wrestlers Dick the Bruiser and Baron von Raschke, and Ricks and her children would put one another in headlocks and deliver flying elbow smashes. Then many nights she'd head out for a marathon walk. She'd leave the house, maybe not to return for hours, venturing off to Kmart or grocery stores miles away. She'd wander by the lake. Or she'd retrace her path the first time she came to Cabrini-Green, hiking the trail back to the West Side, visiting with one family member or another still living there. "Mama, you need to stay in the house some time," Kenosha or Latasha or Earnestine might say to her.

"Who are you talking to? You're not my mama," Ricks would answer in mock anger.

Other nights, Ricks would stay home and barbecue on the ramp, neighbors' children asking if they could have a plate. Of course, she would tell them, just as soon as she fed her own kids. "We all got along like one big family," Ricks liked to say.

The refrain, repeated often by Cabrini-Green residents, was hard to comprehend for outsiders. But neighbors who were like family also hurt one another. Shootings sometimes started while Annie had her after-school classes on the blacktop, and she'd make the children lie flat, herding them in twos and threes back inside the building. Ricks helped her sister get an apartment on the same floor as her, and one day she was visiting when they heard gunshots. Two boys had fired into Annie's apartment window from the open-air walkway. Her son Raqkown was a toddler at the time, and he was playing by the elevators with Kenosha and another girl when it happened. Kenosha huddled the little ones into a corner so they didn't get hurt. But Annie's nephew, who was in his twenties, was inside the apartment, and she found him splayed on the ground, his body outlined in blood and glass. Two shots from two different guns had entered his stomach. The paramedics wouldn't come at first—they didn't think it safe. When her nephew was finally taken to the hospital, he died twice. Both times his heart stopped and he was jolted back to life. "God wouldn't let him die," Ricks would say.

She knew the shooters, along with their brothers and aunts and cousins, and had heard they were jealous of her nephew because he had a job and a new Chevy. One of the gunmen was later shot himself, in an unrelated incident, and he was paralyzed from the waist down. Ricks would see his family carting him around in a wheelchair.

"I didn't care about the shooting after a while, because I got used to the shooting," Ricks would say. "I fear no man, no woman. I fear only God."

FEAR IS WHAT brought Bernard Rose to Cabrini-Green. A Londoner, he wrote and directed the 1992 horror movie *Candyman*, adapting the script from a Clive Barker short story that takes place at a Liverpool public housing estate. When Rose arrived in Chicago, unsure where in the city to set his movie, the guys at the municipal film office told him they knew the place. But they said it was impossible to go there without a police escort. Security detail in tow, Rose roamed one of the Cabrini high-rises; he saw murky stairwells, the caves of apartments blackened by fires, entire floors sealed off and abandoned. "There was something obviously spooky about that," he recalled. He understood that dangerous things happened there, but he also spent time with people like the Rickses, families who were eating dinner, doing homework, and watching television. It felt, too, like everyday life. "The fear around Cabrini was irrational," Rose concluded. It colored every thought people had about the place, blotting out all else. Which for *Candyman*, at least, was perfect. Horror is all about the uncertainty between what's actual and imagined when the danger can't be seen. In the instant before the anticipated attack, in the shadows of the unknown, terror requires that stories of menace flood the blank spaces of the mind. "The old dark house on the hill has always been the standard setting of horror," Rose explained. "But it seemed to me that the big public housing project was the new venue of terror."

Helen, the protagonist of Rose's movie, is a white graduate student in Chicago researching urban legends; she is looking into the myth of the Candyman, a hook-handed apparition who appears when his name is uttered five times. A black cleaning woman at the university, overhearing one of Helen's interviews, says she's from the South Side, not the Near North Side, but she has a friend who has a cousin who's from Cabrini-Green, and *she* says everyone there is scared of the Candyman once it gets dark. This supernatural murderer is rumored to keep a lair inside a block of vacant high-rise apartments. In a film about the hazy boundaries between myth and reality, the fourth-hand information is appropriately vague. It's enough, though, to send Helen to newspaper archives—where she finds reports of other violent crimes at Cabrini-Green attributed to the phantasm— and then into one of the towers.

Helen comes to understand, at least in the movie's mortal sense, that the residents tell tales of the monster as a means to comprehend the incomprehensible brutality of their existence. What else could possibly account for the squalor and isolation and violence in which they live? A societal plot to kill off black people? Utter indifference? Self-inflicted suffering? Why not the Candyman? In this way, the Candyman bears little resemblance to the witch who was said to haunt the barrens along the Ogden Avenue Bridge during Kelvin Cannon's youth. That bit of folklore was evoked to keep children safely inside their homes at the housing development. Candyman was the physical expression of public housing itself as the threat.

But *Candyman* is more about what Cabrini-Green means to those who don't live there. "Just as urban legends are based on the real fears of those who believe in them, so are certain urban locations able to embody fear," the critic Roger Ebert wrote of Cabrini-Green in his three-out-of-four-star review of *Candyman*, in the fall of 1992. The movie's opening images are of Chicago in the colorless gray of early winter as shot from the sky, the contrapuntal piano and organ of Phillip Glass's score adding to the ominous effect. The vision of the city is not of a postindustrial dystopia, with burning cars or darkish hordes. This isn't *Escape from New York* or *Colors*. Instead we look down on the repeating tines of skyscrapers, lifeless in their geometric uniformity and drabness. The aerial shot travels toward Cabrini-Green. The sprawling public housing towers aren't distant outliers; from above, they're not unlike other buildings near them. And that's the point: wherever you are in the inner city, the danger of Cabrini-Green is always in striking distance. The movie references a real-life crime that occurred in Chicago public housing in which a man was able to enter a neighboring high-rise apartment through connected bathroom vanities so cheaply constructed that he simply pushed in the mirrors to create a passageway. Helen goes into the bathroom in her own high-rise condominium and discovers that she, too, can peel off her mirror to enter the apartment behind hers. She learns that her building was originally part of Cabrini-Green. *The projects are coming from inside the house.*

Starting in 1990, Chicago's population began to inch up, the first increase in forty years. The Near North Side saw an influx of four thousand white residents, and the area's median income jumped from a quarter of the city average to twice it during the 1980s. The number of building permits more than quadrupled. Vacant lots that had sold for $30,000 a decade earlier were being snapped up for five times that amount. For the young white professionals who chose to live in the city center, the threat of the ghetto was no longer a story of a far-off place—they now inhabited "transitioning" neighborhoods. Movies such as Martin Scorsese's 1985 comedy *After Hours* tapped into a phobia held by this new Yuppie class—a white office worker finds himself marooned in a yet-to-be-gentrified section of downtown Manhattan. "NONE OF US IS SAFE," one of the endless tabloid headlines blared during the hysteria around the Central Park Jogger case. When Jane Byrne moved into Cabrini-Green, six blocks from her Gold Coast home, she called the housing project "a cancer that can spread to every neighborhood in the city."

In the spring of 1992, the idea that the violence and disorder of the slums could infect a whole city was being demonstrated elsewhere. As *Candyman* was being worked on at a Hollywood studio, rioting erupted in Los Angeles following the not-guilty verdicts of four police officers caught on video beating the black motorist Rodney King. A sound engineer doing a rough mix of the movie switched between the filmed scenes of Cabrini-Green and the televised images of young men running in the streets in South Central. "I'm sorry, I can't do it anymore," he told Rose, as he packed up his bags and fled for home.

In adapting the *Candyman* script, Rose created an elaborate backstory for his killer that tapped into numerous racial tropes. As a corporeal being, Candyman was a gifted portrait artist, the son of a slave at the turn of the nineteenth century whose father earned a fortune after the Civil War by inventing a means to mass-produce shoes. Candyman fell in love with and impregnated one of his subjects, a white woman, and the girl's father hired thugs to lynch him, chasing him to Cabrini-Green (which did not exist), sawing off his painting hand before setting him on fire. In his reincarnated form, Candyman appears in the film gaunt-cheeked and dark-skinned, towering in a fur-lined trench coat, possibly as hell-bent on miscegenation—Helen is a dead ringer for his postbellum beloved—as on murder.

In one of the movie's most frightening scenes, Helen investigates a Cabrini-Green site where legend has it Candyman disemboweled a victim. The setting is the ultimate urban nightmare—a men's public toilet at a giant inner-city housing project. The toilet's tiny, freestanding building sits alone on the concrete plaza between the red highrises, like a decaying shrub overshadowed by giant conifers. It is as much on display for thousands of unseen apartment dwellers as it is abandoned. By herself, of course, Helen enters the public bathroom. Spray-painted threats and indecipherable hieroglyphs cover every surface. She snaps away with a camera for her research. In search of answers, she peers into one stall and gags on the smell. She enters another—the toilets are cracked, overrun, coated in grime. At the last stall she sees graffiti that seems to be a clue. She looks closer, lifting the toilet seat, and recoils. She turns, and a young man is standing there, filling the small space, blocking her exit. He wears a leather trench coat and holds in his hand a metal hook. Three other guys join him, and one of them grabs Helen from behind. "I hear you're looking for Candyman, bitch," the first one says flatly. "Well you found him." Then he takes the blunt end of the hook and knocks Helen unconscious.

The movie at this moment captures some of the true horror in the casualness and banality of the violence at Cabrini-Green. This isn't monster stuff. A guy from the buildings has been appropriating the Candyman myth to prop up his own reputation. Like the shooting into Annie Ricks's apartment that almost killed her nephew twice, it's another routine act amid the city's grinding poverty. The scene ends with the four young men sauntering off, the long camera shot taking in their relaxed stride, the surrounding Cabrini towers, and the downtown Chicago skyline in the near distance.

ONE AFTERNOON, J. R.'s cousin G-Ball and other Castle Crew members noticed a strange white man nearly skipping toward them across the blacktop known as the Killing Field. He appeared as a ghostly apparition, in a tattered old dress, but less Candyman than Casper. The middle-aged man was chubby and balding, wearing thick, cheap glasses, a child's grin smeared across his ruddy face, and waving his arms like a semaphore. "Police," an underage lookout yelled. But anyone with sense could see that this was no cop. What had looked like a dress from a distance was a sort of monk's habit sewn together from dozens of strips of faded blue jeans. The frock was cinched at the waist, and a rosary and a large wooden cross hung from the man's belt. The hobo priest was shouting, "Bud! Buddeee!" lengthening the last syllable, as if they were all lifelong friends. To their surprise, he gave each of them the Gangster Disciples handshake. He was Brother Bill, he said. He was with Catholic Charities, and he would be spending time in the neighborhood.

Brother Bill showed up the next day and most days after that. He'd wander between the buildings, sometimes well past midnight, always stopping by 1117 N. Cleveland to spend time with the Castle Crew, asking about their families or small-talking about the Bulls or the Bears. He never told people to stop selling drugs or to give up their guns. "Hey, Brother Bill," residents would greet him from the ramps and playgrounds. But at the sound of gunfire, whenever there was violence, he hurried to it. He would position himself between the opposing gangs, in the open field and blacktops or alongside one of the towers or in the lobbies, hoping that his presence might stop the shooting. At the very least he wanted to give the gang members an excuse to end a fight. Sometimes bullets zipped by so close that he could hear the vacuum of air as they passed, the metallic taste of fear lingering in his mouth.

Brother Bill believed that he was protected by God's shield, and he could be harmed only if that protection was lifted. As he neared a gunfight, he said a divine voice would instruct him, *Stand here.* It was in those moments, with the surge of adrenaline, the rush of having survived, that he felt a sense of grace: he was fulfilling his purpose, witnessing to save lives. There was a night when he came upon ten teenagers outside one of the white high-rises on Division Street. They were battering another teen with sticks. Brother Bill picked up the boy, shielding him with his body. One of the stick-wielders said, "Brother Bill, you shouldn't do that." Another of the attackers disagreed: "Nah, that's what he's supposed to do."

Years earlier, in 1980, Bill Tomes was deliberating over two job offers, and he dropped into a Ukrainian Catholic Church in Chicago to contemplate his options. While he knelt before the altar, he recalled, the light in the church suddenly changed and the room seemed to spin as if on an axis. His line of sight had nothing to settle on save an illuminated portrait of Christ on the wall above him. And then Tomes heard Jesus: *Love. You are forbidden to do anything other than that.* The disembodied voice commanded, *I'll lead. You follow.* Tomes had no doubt that the son of God had spoken to him. He'd experienced a miracle. Even the Catholic Church would later recognize that he'd undergone a genuine locution, a private revelation. But he didn't take himself seriously enough to believe he had a higher calling. He assumed Jesus had made a mistake. Not long after, while lying awake in bed, he felt Christ standing before him, uttering, *You will have me. You will be poor.* Bill responded, "Poor, wow, not too tempting. Yuck. Such a deal." Christ went on, *You must forgive everyone everything.*

Eventually, Bill stopped resisting. He took a vow of poverty and volunteered at one of the poorest parishes in all of Chicago, Saint Malachy's. The Near West Side parish had over the years catered to Irish families, Italians, and Slavs. It now served the residents of the Henry Horner Homes, a public housing complex of nineteen buildings. Alex Kotlowitz's 1991 nonfiction bestseller, *There Are No Children Here*, was about two brothers, Lafeyette and Pharoah Rivers, living in the Horner Homes with their mother, LaJoe, and their siblings. "Twice in May, LaJoe herded the children into the hallway, where they crouched against the walls to avoid stray bullets," Kotlowitz writes. "Pharoah's stutter worsened, so that he barely talked and stayed mostly by himself. He continued to shake whenever he heard a loud noise. Lafeyette told his mother, 'Mama, if we don't get away someone's gonna end up dead. I feel it.'" A great number of the young men Tomes came to know at Horner were members of gangs. He found that he liked talking to them. He'd labor at the parish in the morning and walk the housing development from noon until late at night. Sometimes it was difficult to forgive murderers, people who could fire indiscriminately into a crowd or sell drugs to a young mother at her wit's end. But forgiveness, as Christ had instructed him, was the key to the work.

By the eighties, Chicago's Cardinal Joseph Bernardin felt the Catholic Church needed to do more to address the violence afflicting the city's parishes, and he asked Tomes to carry out his work with gang members in other parts of the city as well. Tomes had no calling for the priesthood, so he couldn't be a father. Officially he would be a consultant with Catholic Charities, earning a small stipend, with the church allowing him to go by "Brother Bill," a brother in the sense of his union with all people. The place associated most readily with violence in Chicago was Cabrini-Green, so that's where Brother Bill was sent. He modeled his habit of stitched denim after the sewn rags worn by Saint Francis of Assisi. It looked like the oldest pair of jeans you owned, the denim turned white with wear, tessellated sections gone to holes and then patched, the many parts pieced together like a Frankenstein monster. He could be a holy man or a homeless man. He wore a homemade scapular as well, and his frock had a hood that he never donned, since he felt it looked too Ku Klux Klan. The archdiocese approved the outfit, almost as a kindness to the gang members. Tomes stood out even late at night at public housing; no one would shoot him by accident.

The longer Tomes did this work, the less comfortable he was leaving the land. He'd stay at Cabrini-Green or Henry Horner or Rockwell Gardens until 2:00 a.m., waking for morning mass, and then repeating the cycle day after day. Over the years he saved the programs from each of the funerals he'd attended as Brother Bill—his stack growing from 10 to 80 to 130. He rarely slept, and he berated himself when someone was hurt in his absence, as if he could have done more. He should have been in the stairwell, in the elevator, on Oak Street, the rowhouses, Stanton Park, on the blacktop, by the El tracks, in front of the Rock, outside the Lower North Center.

J. R. FLEMING

THE SUMMER OF the first Bulls championship, in 1991, was miserable with murder: 121 homicides in August alone, the most for any month in Chicago's history, with the official count for the year surpassing 920. More days than not were peaceful at Cabrini, but sniper fire was an especially effective weapon of terror. Employees of Commonwealth Edison refused to do maintenance at a nearby substation, fearing that from inside a cherry picker fifty feet high they'd appear as targets. Some of the workers weather-sealing the concrete exteriors of the Cabrini high-rises wore bulletproof vests. When J. R. Fleming stepped outside, he would look up to scan the top floors of the towers around him as if checking for bad weather. If he saw white sheets covering the windows, he figured it was probably safe to hustle across the blacktop. But if he didn't see the sheet, then it was possible someone behind the window was scanning the grounds with a rifle in hand. The thing was, J. R. never knew whether or not he was in the crosshairs, whether the bolt of lightning would suddenly strike.

The high-rises forming the south and north boundaries of the Killing Field fired on the towers lining the east and west borders, and vice versa. In an internal memo, a housing manager at Cabrini told the CHA that snipers in 1157–1159 N. Cleveland had shot at least ten people from 500–502 W. Oak. Every building had some Vietnam veterans, but even untrained guys would go up there to feel the power of aiming the big gun, deciding whether the tiny objects moving below would live or die. "I live in a neighborhood I am sure you've heard of, Cabrini-Green Housing Development," a longtime tenant named Mrs.

Henry Johns wrote in a letter to President George H. W. Bush. "The innocent residents of this community cannot walk the streets without fearing for their lives," she explained. "I believe in your foreign policy. The United States must take a major role in world affairs, but to be effective in that role we must also set an example here at home. . . . My solution would be to do what was been done in the Persian Gulf. Send in the troops and get rid of the opposition and ammunition."

One night when J. R. and his friends were leaving a liquor store on Larrabee, they heard the percussive blows before they saw the little eruptions on the ground around them. It took a moment for J. R. to realize that someone was shooting at them. They dropped their bottles, fell flat on their chests, and crawled military style on elbows and knees back into the store. Once inside, they patted themselves to check if they'd been hit. They were fine. But now they were trapped. They gave their handguns to the clerk, who stashed them away and did the only thing possible: called the police. The two cops who showed up cursed J. R. and the guys with him, resentful that they were risking their lives to deal with this nonsense. Thirty minutes later, four paddy wagons parked front to back to form a wall across Larrabee, and J. R. and his friends crossed the street using the cars as cover.

J. R.'s godmother was one of the women who ran his high-rise. LueElla Edwards had left Cabrini-Green for a time, moving her family to the Harold Ickes Homes, a public housing development on the South Side, but she decided her children were in graver danger there. At least at Cabrini she knew everyone. As far as the young men hanging out in front of her high-rise, she demanded of them a kind of tax. When J. R. entered the lobby of his building, Edwards would charge out of the tenant council office and say she'd checked the sign-up sheet and he hadn't put in his hours of service. Like all the other guys, he was expected to sweep and mop the ramps, monitor the building's first-floor computer club, and volunteer for Edwards's Take Our Daughters to Work Club. They also chaperoned trips for the children from the building to baseball games or Six Flags Great America, and they were sent across the field to help with Holy Family's Boys in the Hood youth group. There was a weight bench in the 1017 lobby, and J. R. might be waiting to get in his reps when he was ushered into a tenant council meeting, him slouching in the back of the community room, listening abstractedly to the women talk about the city's plans to kick them all off the land at Cabrini. "I had money then. I'm my own boss," J. R. would say. "The end of Cabrini wasn't impacting me."

It was LueElla Edwards who introduced J. R. to Brother Jim, another lay member of the Catholic Church who'd joined Brother Bill in walking the city's public housing developments. Twenty years younger than Bill Tomes, Jim Fogarty also wore a habit of stitched denim. Tall and athletic, he'd take it off when he joined pickup basketball games, realizing that J. R. and other young guys wouldn't really try if he had on the robes. He was a student at a Chicago seminary when Bill showed up one day, describing how he stopped gunfire and asking if anyone cared to join him. Later in life, Fogarty would parse more deliberately the line between the real and the symbolic, between meaning and myth. But back then it looked to him as though Brother Bill had emerged from the pages of *Lives of the Saints*. Fogarty believed that special people in special places could have divine experiences, and he was eager to see how the Lord worked through this odd man.

For the media covering the urban crisis of the 1990s, Brother Bill was irresistible as well—a white man who looked like he'd stepped out of the Middle Ages claiming that God had sent him to the most infamous public housing project in the country. *Time* magazine did a feature on him, and in August 1992, a camera crew from the national news show *Eye on America* followed Brother Bill around Cabrini-Green. Here was "the most dangerous patch of blacktop in America," the host intoned, and Brother Bill was the "street gang missionary" who had survived some thirty near hits and saved hundreds of lives. The reporters from the program interviewed LueElla Edwards, who said her fifteen-year-old daughter, Laquanda, had begged her to move them away from Cabrini-Green. Edwards said she worried every time her children went out to play, but all she could do was pray. With Brother Bill fitted with a microphone, cameras trailed him as he lurked in the shadows, waiting for the gunfire that would set him in motion. A shooting did occur on the land that night, and Tomes ran to it. Sniper fire had struck a girl in the back of the head as she walked on Larrabee near Holy Family. Bending over the victim, with the cameras rolling, Brother Bill looked into the girl's face. It was Laquanda Edwards, interviewed by the news crew only hours before. She had been on her way to the corner store to buy milk. Bill wept over the body. Behind him, J. R. skulked back and forth, set on revenge, a .357 Magnum stuffed into his sweatpants, as Brother Jim talked him down.

Brother Jim helped J. R. secure a peddling license, and he found work for many others, although most of the jobs didn't stick. He even offered J. R. a way out of Cabrini-Green. A group that included Judge Reinhold, the actor from *Beverly Hills Cop* and *Fast Times at Ridgemont High*, bought the film rights to Brother Bill's life. The production faltered, and Brother Jim arranged for J. R. to provide the Hollywood guys with some vérité footage of life on the ground at Cabrini. J. R. already owned the video equipment, and he discovered that he loved the process of filming and editing, creating a story out of a thousand stray moments. He'd capture his friends as they gathered in front of their high-rise or crossed Larrabee on their way to the mall or hung out inside Sammy's Red Hots, across from Atrium Village beside the El tracks. He sometimes stage-managed, ordering everyone back inside the hot dog stand so he could shoot their exits all over again, instructing them to walk more naturally. He started filming young rappers from Cabrini, and suddenly every guy who wanted to show off his rhyming skills was seeking him out for an audition. The team from Hollywood liked what it saw. Brother Jim suggested J. R.

do like Preach at the end of *Cooley High* and quit Cabrini-Green for the West Coast movie industry. J. R. had savings from his peddling, and he could ask Reinhold for a starter job in film. It was appealing, the idea of the total reinvention.

But J. R. couldn't leave. He told Jim that he already knew all about the Bloods and Crips from what he'd heard in West Coast rap and seen in movies like *Colors*, *Boyz n the Hood*, and *South Central*. The LA gangs had become as infamous, as much of an urban bogeyman, as Cabrini-Green. If J. R. lived out west, he half-joked, he wouldn't be safe wearing any color other than orange. He was less afraid of the violence, though, than of the unknown of a new place. He'd rather stick it out with what was familiar, where his friends and family lived. For better and for worse, he'd announce, "I am Cabrini-Green."

11

Dantrell Davis Way

She was twelve, and he was fifteen and to Annette seemed somehow better than the other guys stationed in the lobby. Sharper dressed, funnier, finer. That's how she knew it was love. In 1982, Kelvin Davis lived on the West Side with his mother, but he showed up each day at Cabrini-Green. K-Mac, they called him. There were lots of boys and also men who tried to talk to Annette Freeman. She was a tomboy, short and cute, with close-cropped hair. It was Kelvin, though, who understood her suffering. Annette's father died the previous year. Shot dead on the South Side. She'd been running away from her mother's apartment in 500 W. Oak ever since. Janice Freeman would drink, beat Annette, and then beg forgiveness after sobering up, only to wake Annette later that night with more blows. So Annette left home. "It's time to man up," she told herself. She refused to sell her body, choosing to sell weed instead. She got a job selling newspapers, too, hawking them on Chicago Avenue, State, and Michigan, the only girl out there with all the boys. Some nights Annette slept at her grandmother's house, more than a hundred blocks south, but each morning she hopped on the El to come back to her building. Kelvin would be there, and he'd call her the "lost child" and make her laugh through her pain.

Annette was fourteen, in 1984, when she learned she was pregnant. It seemed to her like the end of the world. Among her friends

from the building, she was the first to get pregnant. She was humili-
ated. Her belly sticking out, announcing that she was having sex. So
much for manning up. She had to quit the job selling newspapers.
"Here I was a child and homeless and about to have a child of my
own," she recalled. She wondered what kind of mother she could
possibly be. But Kelvin kept telling her it was going to be all right.
"We're going to be okay, Ann," he repeated. They named the baby
after two of Kelvin's uncles—Dantrell Tremaine Davis. Annette
called him Danny.

Annette had been abandoned as a child, left for several years
with a woman on the South Side she didn't know, and because she
was still a ward of the state, the Department of Children and Family
Services wanted her to sign over parental rights. A baby was a burden, no doubt, but she wouldn't give Danny up. No way. She wouldn't
even leave him alone with other people, not after what had happened
to her. The two of them went everywhere together. She took him with
her to Cabrini, and they would stroll downtown or to the Lincoln
Park Zoo or to the beaches on Oak Street and North Avenue. They'd
walk to the Milk Duds factory to get candies, and wander Old Town,
looking in the bike shop and the Ripley's Believe It or Not! Museum.
Kelvin had other children with other girls, but he'd join them sometimes, the three of them like a family. Annette and Danny slept at her
grandmother's, or at Kelvin's mother's place, or in the elevator of 500
W. Oak. Occasionally she rode the train with Danny, going from one
end of the line to the other until morning.

When Annette turned eighteen, and Danny was three, she filled
out an application for her own apartment at Cabrini-Green. She would
never forget the day she got it—February 9, 1989. The unit was on
the sixth floor of 502 W. Oak, the twin connected to her mother's
building, and moving in was her freedom. It was liberation not only
from the uncertainty of homelessness, the cold, and strange men,
but also from all the naysayers who told her she wouldn't accomplish
anything, that she'd probably be dead by age eighteen and couldn't

take care of a child. Her uncle Henry, a bus driver, helped her with furniture. An official from family services checking on her saw that the apartment was clean, that she was providing for her son, and Annette was granted guardianship over both Dantrell and, for the first time, her own self.

Annette made sure Danny played baseball at Cabrini and boxed in the Seward Park field house. She never let him miss a day of school. No way would Danny grow up to be in a gang and sell drugs. She sold drugs, but that was to keep Danny in the shoes and clothes she couldn't afford on a public aid check alone. She took one of the rotating shifts outside her building. With so many others needing the work, she barely got any hours. When Dantrell acted up and was sent to a kindergarten for children with behavioral problems, Annette's friends told her she was lucky. She could claim an SSI, supplemental security income, for someone with a disability. "I'm not going to claim shit extra," she said, "because my son is intelligent." He was hardheaded, that's all. He didn't listen to anyone but his mother and father. She rode the bus with him to the school each day, making sure he was taken seriously as a student. She lobbied to get him reassigned to a school at Cabrini for first grade.

In the spring of that kindergarten year, in 1992, with Dantrell asleep on their couch, Annette left him for a shift outside the building. At 4:00 a.m., a woman who lived next door heard Dantrell's screams before she smelled the smoke. A man from the building kicked in the door. A cigarette by the couch must have started the fire. An ambulance rushed Dantrell to the emergency room while firefighters put out the blaze and the police located Annette. The burns covered half of Dantrell's body, mostly on his left side, over his stomach and arms and face. His cheeks were polka-dotted in pink where the skin seared, his fingertips hardened like wood. He spent the next few weeks in the hospital, Annette and Kelvin by his side. After the skin grafts, you could see the burns, but they were much less visible.

They got another apartment in 502 W. Oak. Annette was ordered to take parenting classes. She was happy to do it. She was told she couldn't leave him alone again. She understood. After weeks in the hospital, Dantrell was ready to start first grade at Jenner. That was perfect. The elementary school was right outside their high-rise, not a hundred feet from their front door. Although he still couldn't read or print out letters, his new teacher said it would be only a matter of time, and Danny believed him. At the start of the school day he sat transfixed as the teacher read stories to the class. He won second place in a boxing tournament, and his coach thought his wide smile electric. Annette felt they were doing good. They'd both overcome so much. Dantrell told her all the time that he loved her. He'd repeat it like a song—*I love, love, love you.* He was only seven, but he had an old soul. "He didn't give up on me, even when I gave up on myself," Annette would say. "Not my little soldier, not my best friend."

Then on September 13, a week into the school year, Kelvin Davis died. He was twenty-seven, his body discovered in bed at his mother's West Side home. Annette would blame asthma; official reports stated that he choked in his sleep, the likely result of a drug overdose, either heroin or methadone. His rap sheet was long on petty crimes—possession, assault, shoplifting. He had nine children total, but Annette said he'd been a good father to Danny, and he was the only man she ever loved. They'd been together ten years, almost half her life, and the death made her want to drink herself senseless or die or hurt other people. She held it together for Danny. "I miss him, too," she'd tell her son. "But it's only us now."

In October, a couple of weeks after the funeral, Kelvin's sister dropped by to check on them. The night was cold, and as they walked Dantrell's aunt back to the bus stop Annette looked down at her son trudging silently between them, his face twisted in worry. "Man, Ma, I don't want to get shot," he said. It was a peculiar thing for him to say. He'd never fretted over the violence before. Their building was Vice Lords, and the towers flanking them were controlled by Gangster Disciples. When Annette was in high school, she and her friends started their walk to Lincoln Park each morning by counting down before breaking into a sprint, with guys from the surrounding high-rises dropping things on them or streaming outside with golf clubs and giving chase. Annette trained Dantrell from an early age to stay clear of the blacktop, to walk away when fights started. At the sound of a gun blast, he knew to fall on the ground. He knew the drill. She told him not to worry. "Don't I always got you, Danny?" she said, pulling him into her waist. "I'll take care of you."

The next morning, October 13, 1992, was a Tuesday, and Dantrell loafed in his room, watching cartoons while Annette dressed. That wasn't like him; he usually set the pace, beating her to the door. "You're not going to school today?" Annette teased. He said he wanted to stay home and spend the day with her. But skipping school wasn't part of their plan. When he still didn't budge, she agreed reluctantly to walk him downstairs. They came out of the building, and Annette waited for him to cross the street to Jenner. Several teachers were outside greeting students. Parents in shiny yellow vests worked as crossing guards. Two police officers idled in a squad car at the corner. Annette pointed to Danny's friends. "There goes Doo-Doo," she told him. But he didn't move. "Didn't I tell you to go over by the school?" She urged him on with a flick of her hands, irritated in the way parents get when their children are slow to react. "Do something," she snapped.

"I am doing something, Ma."

Then the 9:00 a.m. school bell sounded. A moment later, Annette heard the first of the shots. She dropped to the ground and yelled for Dantrell to do the same. When she looked over, she saw him already on his stomach, and a bit of pride blossomed in her. "That's because of his learnings," she told herself. She had prepared him; she was keeping him safe. The shooting stopped, and Annette crawled over to her son. "All right, let's get up. Let's go in the building," she said. But he couldn't get up. Bystanders said she screamed, "Please, baby, don't die!" and, "Please, come and get my baby! Please hurry!" She didn't remember any of that. A half hour later, at Children's Memorial Hospital, Dantrell Davis was pronounced dead. A bullet had entered the left side of his head.

DANTRELL DAVIS WAS the 782nd person murdered in Chicago in 1992. That was in October, and by year's end the official count would reach 943—more homicides than in any year in city history except 1974, when Chicago had 970 killings and almost 600,000 more residents. Much of the violence was blamed on gangs and drugs. It was the height of crack addiction, with dealers and buyers and hard-core users congregating in vice districts like Cabrini-Green. Dantrell was also the seventy-fourth child to be killed that year. He wasn't even singular as far as victims from Jenner Elementary. Three months earlier, it had been Laquanda Edwards, buying milk on Larrabee Street, struck down by a sniper; in March, second-grader Anthony Felton was standing almost exactly where Dantrell died when thirteen-year-old Rodnell Dennis fired blindly into a crowd and killed the younger boy. "There were lots of Dantrell Davises before Dantrell Davis," Dolores Wilson's daughter Cheryl said.

It is a trick of fate, the perfect storm of circumstance, that raises any one tragedy into something other than an installment in a city's daily accounting of horrors. The shameful rule is that bystanders are gunned down, boys are shot by police, children murder other children, and what's shocking today elicits no extended public outcry. Dantrell Davis proved to be the rare exception. "Dantrell will go down in the history books of this country," Vince Lane, the CHA head, announced to reporters in the days after the shooting. "Dantrell has gotten the world's attention." That he became, as the editor of the *Sun-Times* declared, "the symbol that Emmett Till was," was due in part to the circumstances of his death. He stood only a couple dozen steps from his school's entrance, first thing in the morning, surrounded by teachers and police. He and his mother were side by side. He represented any seven-year-old starting not merely his day but a lifetime of untold experiences. That he was vulnerable even then captured the growing feeling that the city had finally broken. It had lost its capacity for mercy. Cardinal Joseph Bernardin said that Dantrell's death underscored "the social pathology that we have refused to acknowledge."

Still, Dantrell's brief life would likely have been forgotten if he had not been from Cabrini-Green. Dantrell Davis and Cabrini-Green were talked about together as the manifestation of the crisis of inner-city America. In the extended coverage, the media outlets each ran long treatises about the woeful history of the development. "Cabrini-Green was born in Little Hell," wrote the *Tribune*. "The name is still sadly apt." On national news shows, Cabrini-Green was "Fort Cabrini," compared with Beirut, Sarajevo, and Somalia. "A public housing hell-hole of poverty and crime," reported the *CBS Evening News*. One of the hundreds of newspaper articles on the shooting stated, "Now the very name 'Cabrini' is synonymous with what many people fear most about urban America: rage and hopelessness spiraling out of control." The reactions made sense, almost, when considering that Dantrell Davis and *Candyman* were complete contemporaries. On the last night of Dantrell's life, as he walked with his mother and aunt, the horror movie set at Cabrini-Green was premiering at the Chicago International Film Festival just four blocks away.

On the morning that Dantrell was shot, at roughly the exact same time, three hundred officers were raiding Altgeld Gardens, the 190-acre public housing village on the city's southern edge. A young Barack Obama did community organizing there for three years in the 1980s, trying to get asbestos removed from the buildings and an after-school program started for teenagers. He found that the absence of any of the social order he'd witnessed in the slums of Jakarta as a child "made a place like Altgeld so desperate." Obama wrote in his memoir, "It wasn't as bad as Chicago's high-rise projects yet, the Robert Taylors and Cabrini Greens, with their ink-black stairwells and urine-stained lobbies and random shootings." He was right about the "yet." With the closing of the steel mills that were the reason for building Altgeld way out there in the first place, the development's thousands of residents were cut off from both the city and from jobs. It was now part of what was colloquially known as the Wild 100s, the dangerous and depressed three-digit streets of the distant South Side. The officers conducting the early-morning raid turned up twenty guns, a small cache of drugs, and portable two-way radios that officials said drug dealers were using to coordinate their operations. Thirty-five people at the development were arrested.

But Altgeld, sixteen miles from the city center and a world away, was hardly news. When Lane got word of the shooting on the Near North Side, he sped across town. He knew that Dantrell's death would be a sensation. "It was Cabrini-Green!" he explained more than two decades later with a cry. "Nobody cared about the other projects. Cabrini sat in the heart of the city. The world knew about Cabrini-Green." By the time he arrived, mothers and grandmothers had already collected their children from Jenner. Studentless teachers huddled together in a daze. Dozens of police cars with their whirling blue lights were parked at uneven angles. A traffic jam of news trailers had formed along the streets, while reporters and city officials ran clumsily between buildings with their heads ducked, fearing additional fire. A spent rifle cartridge was found in one of the high-rises facing 500–502 W. Oak, and sixty officers charged into the building. They determined that the shot likely came from a window in an abandoned tenth-floor apartment in 1157 N. Cleveland, about three hundred yards from where Dantrell fell. By noon the next day, the police had arrested Anthony Garrett, a thirty-three-year-old lifetime Cabrini resident.

Richard M. Daley was forty-nine and had been mayor for less than three years when Dantrell Davis was shot. He had defeated Harold Washington's successor, Eugene Sawyer, in a special election in 1989, and two years later was reelected to his first full term, his 63 percent of the vote in the Democratic primary exceeding even his father's best electoral showing. After twenty-one years of Old Man Daley, five different mayors ran Chicago in a little over a decade. The younger Daley seemed like a kind of stability, especially after the turmoil of the Council Wars that beset Washington's mayoralty. And unlike Jane Byrne, who made several racial missteps early in her tenure, Daley appointed minority members to his cabinet and to head many of the city's departments. He formed alliances with leaders in the city's black and Latino communities, and he became the first mayor to march at the head of the city's Gay and Lesbian Pride Parade. Prior to becoming mayor, Daley had served two decades as an Illinois state senator and the Cook County state's attorney. But he distanced himself from his father's Democratic machine, whose faithful still blamed him for splitting the white electorate with Byrne in 1983, thus allowing Washington to win office.

In the thirteen years between the two Mayor Daleys, the party machinery had actually been rendered somewhat obsolete. The Shakman decrees that were signed into law in entirety in 1983 illegalized much of the patronage system that bartered city and state positions in exchange for political support. By the 1990s, political bonds were sealed not through the doling out of city jobs but through the dispensing of city contracts. The younger Daley proved adept at the new system, forging ties with Republicans in the suburbs, where many city companies had relocated. He began filling vacancies on the city council with aldermen loyal not to the Democratic Organization of Cook County but to him. He reordered city bureaucracies, made sure the White Sox remained in Chicago, and invested heavily in a resurgent downtown. He also benefited greatly from an improved economy. After forty years of population declines, a net loss of 837,000 people, Chicago began to add residents again as Daley took office. His was the new gleaming Chicago of Michael Jordan and Oprah and Charlie Trotter. Daley himself would move, in 1993, from the old sod of Bridgeport to the luxury Central Station, a former rail yard just south of the Loop that had been one of the primary targets for rede-velopment in the Chicago 21 plan.

While Daley had managed the city up to then with a low-key style, he'd yet to show much leadership on the growing problem of violent crime. The number of murders in the city rose each year he'd been in office—742 in 1989, then 851 in 1990, and 928 in 1991. Throughout his tenure, Daley would be mocked and also praised for his sometimes garbled speech—a sign, trusted aides would say, of his too-fast-for-his-mouth thinking. In a press conference, he denounced "drive-by shootalongs." At another he proclaimed, "The more killings and homicides you have, the more havoc it prevents." At the time of Dantrell Davis's murder, Daley was visiting his daughter at Con-necticut College, and he didn't gauge the political significance of this death. In a record bloody year, Davis seemed another painful loss, a tragedy, but not an all-hands-on-deck civic emergency, or even cause for a public comment. Jane Byrne figured out early on the space that Cabrini-Green held in the civic psyche, and she'd turned a shooting spree there into the most prominent political moment of her career. By contrast, Daley's ongoing absence was compared with the schmoozing of Los Angeles's police chief, Daryl Gates, at a Brentwood cocktail party as rioters tore their way through his city earlier in the year. The *Sun-Times* wrote of the Chicago mayor, "This city cried out for leadership last week, and the biggest chair of all was empty." Daley slammed his critics, saying that family was of the utmost importance, reminding more than a few Chicagoans of his father's response to reports that he'd helped the younger Daley and his brothers get lucra-tive city contracts and court appointments. "If I can't help my sons, then they can kiss my ass," Richard J. Daley had said in 1973. "If a man can't put his arms around his sons, then what kind of world are we living in?" But rather than return to Chicago after visiting his daughter, Richard M. Daley traveled to Hyannis Port, Massachusetts, to join the Kennedys at a golf tournament.

With Daley gone, Vince Lane stepped onto the public stage, announcing into a bouquet of microphones that he would do whatever was needed to turn the infamous public housing project into a "normal neighborhood." He was ready to send in a thousand additional state and federal law enforcement officers to sweep every building. The tower from which Dantrell's killer had fired the fatal shot already had an official vacancy rate of over 75 percent. Lane declared that the building would be permanently closed—"And I don't mean with plywood; I mean with masonry." He called on the National Guard to police Cabrini-Green, demanding that armed troops form a perimeter around the development's twenty-three towers and all six hundred of the rowhouses.

Daley decided it was time to cut short his trip. Lane couldn't be the one making all the pronouncements. The mayor announced a meeting with his advisers for Monday morning, six days after Davis's killing. Then the emergency meeting with twenty of his top officials was moved up to Sunday. At the press conference afterward, Daley appeared sunburned—his aides said from activities the mayor did in Chicago, before his golf trip. Daley presented an eleven-point plan for Cabrini-Green. He adopted several of Lane's impromptu proposals as part of his own strategy, saying that he would shutter up to four Cabrini high-rises and outfit the remaining buildings with metal detectors and one-way turnstiles. He'd place armed guards in the lobbies. His agencies would expedite evictions, putting together a team of lawyers who'd work pro bono to remove drug dealers and gangbangers. The city would pay 270 off-duty officers to sweep every building in the project of weapons, at a cost of upward of $500,000. Adopting the language of the War on Drugs, he promised Chicagoans that he would take their city back from gangsters and make it safe for all law-abiding citizens. "We cannot surrender," Daley pledged. "We refuse to stand by in a city where a seven-year-old can't walk from his home to his school without fear of death." Echoing Sean Connery's character in *The Untouchables*, who promised to take down

Al Capone the "Chicago way," the mayor said he would do what was necessary to eliminate the gangs in areas like Cabrini-Green: "We have to have a war here, and we have to go after them the same way they go after innocent people."

AT THE FUNERAL service, four days after his death, Dantrell Davis was outfitted in a white tuxedo with a powder-blue cummerbund, a bow tie, and a white leather cap that covered the damage to his head. The pastor described Danny as a warrior, a little David toiling in a community assailed by Goliaths. A soloist sang the lament that ends the Cabrini-Green movie *Cooley High*—"It's so hard to say goodbye to yesterday." People heard the song and crumpled. At the cemetery, Annette Freeman kissed a red carnation and tossed it onto her son's casket. As the coffin descended into the ground, she yelled that Danny had been the only one left who loved her.

Annette's mother, Janice Freeman, had moved to Cabrini-Green for safety. She was escaping an uncle who had tried to kill her—a different uncle than Jerome Freeman, known as King Shorty. He was the leader of the Black Disciples, a Chicago gang with thousands of members. He'd joined the gang as a boy, running errands for its founder, David Barksdale. When Barksdale created the Black Gangster Disciples Nation, he became King David, and his Folks alliance adopted as its emblem the six-point star—the Star of David. The People alliance used the five-point star as its symbol. During a Blackstone Rangers ambush in 1970, Barksdale was shot in the abdomen. He survived, but the complications persisted, and four years later he died of kidney failure. The Disciples Nation died along with him. Shorty Freeman took over the Black Disciples, and Larry Hoover the Gangster Disciples.

At the time of his grandnephew's murder, Shorty Freeman was two years into a twenty-eight-year sentence for drug trafficking. Hoover and Jeff Fort, of the Blackstone Rangers, were also serving long sentences in prison, all three leaders reportedly running their street operations from behind bars. At Stateville, Freeman held regular meetings in the prison chapel, and Annette was driven to the penitentiary to talk to him about Dantrell's accused killer, Anthony Garrett. In a recorded confession, Garrett admitted to standing on a bathtub in an empty apartment in 1157 N. Cleveland and firing an AR-15 rifle at a group of teenagers outside Annette's building. The weapon lacked a scope, and he'd missed his target and struck Dantrell by mistake. The next day, Garrett claimed he was innocent and that the confession had been coerced. He said he'd been held in isolation for five hours and beaten until he agreed to whatever police told him to say—not an implausible charge in Chicago. Revelations were emerging then about a police unit under Commander Jon Burge that had tortured more than a hundred black suspects over the previous twenty years, beating them with phone books, suffocating them, and shocking them with electrical devices on their genitals and in their rectums.

Garrett was known around Cabrini-Green as Quabine. A former army specialist fourth class trained in marksmanship, he'd racked up nine convictions from the time he was eighteen, six of them weapons related. In 1981, on the same day that Mayor Byrne announced her plans to move to Cabrini-Green, he'd been arrested for killing another resident in a high-rise stairwell. Although he would be found not guilty of the crime, police said he admitted to the shooting and bragged about being a Cobra Stone leader in charge of multiple buildings. Months before Dantrell was killed, during one of Vince Lane's door-to-door sweeps, police discovered a loaded gun in his Cabrini-Green apartment. A judge later ruled that officers illegally seized the weapon, since the CHA raids violated all sorts of protections from unlawful searches, and Garrett was released.

Like many people at Cabrini-Green, Garrett wasn't just one thing. He umpired baseball games at the housing development, volunteering for free for three years until he was able to earn $8 an hour for his time. He was Dantrell Davis's Little League coach, and Annette knew him. "He used to like on me and everything, but I wasn't leaving

Danny's daddy," she would say. "I was very loyal." Garrett makes an appearance in Daniel Coyle's *Hardball*, a book documenting a Little League season at Cabrini-Green over the months before Dantrell's murder. Coyle writes that Garrett was the best umpire the teams saw: "A muscular man with a coal-black biblical beard, he approached the game with extreme seriousness, bellowing 'ball' and 'strike' with an undisputed authority." "'I like working with kids,'" Garrett tells the coaches. "'I ain't always been straight in my life, but if I can keep a few of them out of trouble, then I done my job.'" One of the league's corporate sponsors was so impressed by Garrett that he helped him land a maintenance job a short walk from Cabrini at the posh East Bank Club ("a membership to East Bank Club is an invitation to live well").

The rifle that killed Dantrell was never found, and some police officers at Cabrini assumed Garrett was the fall guy for a heater case that needed an immediate culprit. But neither Annette nor Shorty Freeman were concerned with whether Garrett was innocent or guilty. They both figured that Quabine killed Dantrell by accident. Plus, Annette was insane with grief. And Shorty knew people were waiting to see his reaction. The moment required that he be a big man. Because Dantrell was kin, his murder had the potential to set off a series of deadly reprisals. No one wanted that. Not unlike Mayor Daley or Vince Lane, the gang leaders needed to demonstrate that they could impose order on a city that more frequently seemed without any. Organized crime was one thing, disorganized crime something else entirely.

Shorty let it be known among his associates that there would be no revenge. Too many people were already getting hurt under his name. But when Garrett began his hundred-year sentence in a Stateville cell house where Shorty called the shots, he didn't stop guys from delivering some private retribution.

EXACTLY A WEEK after Davis's shooting, also just after 9:00 a.m., hundreds of Chicago police officers and city officials rolled into

Cabrini-Green, joined by officers from the CHA's security force, the Cook County sheriff's department, the DEA, ATF, and FBI. The police brought in their blue militarized mobile command vehicle, planting it in front of the red high-rises. Platoons of armed officers went floor to floor in two of the high-rises, searching every room in every apartment. Another hundred recruits from the police academy joined the sweeps. Men and teenage boys were flushed from the buildings and searched repeatedly. One resident who joined a lawsuit against the CHA complained that he was searched seven times in one hour, and not just pat downs but police going inside his underwear; his two-year-old son was frisked. The officers turned up just one firearm—a .45 caliber replica of a gun unable to fire ammunition.

In addition to the show of force, Mayor Daley sent in a hundred street and sanitation workers. They put in floodlights, laid out poison traps for rats, filled potholes, and scoured the project for abandoned cars to tow. Workers applied fresh coats of paint to graffiti-covered walls. Trimmers lopped off branches, and electricians repaired wires. Carpenters nailed plywood over vacant units that had remained accessible for years. Thirty men who had struck deals with the county to exchange jail time for community service cleared the grounds of trash and weeds. Teams installed metal detectors and guard posts, while crews of counselors from the city's Department of Human Services met with residents who'd been forced from their homes.

Reporters were allowed to embed with roving teams of social workers, clergy, CHA officials, and armed policemen. Ethan Michaeli, the publisher of *Residents' Journal*, a newspaper by and for public housing tenants, was then a cub reporter with the *Defender* and joined one of these tours. The sense was of an emergency relief force showing up in a disaster zone—"We're the government, we're here to help you." At one high-rise apartment, a well-dressed young woman who answered the knocking was asked if she needed any assistance. She was confused: "With what?" Staring at the crowd outside her door, however, she couldn't help but mind her manners. Would they

like to come in? Michaeli said the reporters, eager to experience an
actual Cabrini-Green public housing unit, crammed into the woman's
home. But they were disappointed. The living room and kitchen were
simple but clean. A shag carpet covered the floor and long, white
curtains decorated the windows. A neatly attired toddler stared up at
them in wonder. When Michaeli lingered later on one of the ramps,
surprised by the vista of downtown, someone from the CHA ran up to
him, breathless. "I thought we lost you," she cried.

Residents of Cabrini-Green had been pleading for regular maintenance, for new light bulbs in the stairwells and roach fumigation
and elevator repairs. In the name of Dantrell Davis, all the stored-up
service had finally arrived at once like something out of a fable. With
the truckloads of heavy equipment, the workmen bustling about, and
the sounds of the chain saws and welders, it was as if a Hollywood
crew were creating the set of an imaginary public housing project
from scratch. It also seemed like a bizarre reenactment of events from
1970 and from 1981, after the killings of Severin and Rizzato and the
arrival of Mayor Byrne. The city's entire apparatus appeared to be
focused again on a seventy-acre patch of the Near North Side, home
to only 8 percent of the CHA's population. There were miles of land
on the West Side that still hadn't been rebuilt since the fires in 1968.
At the Henry Horner Homes, half of the 1,800 units were vacant, up
from just 2 percent only a decade before, as the CHA wasn't both-
ering to repair or refill empty apartments. Seven other public housing developments in the city had rates of violent crime that exceeded
Cabrini's, and the neighborhood including the Robert Taylor Homes
and Stateway Gardens had the highest proportion of murders in Chicago. "They're putting a lot of manpower in Cabrini now, but other
neighborhoods need the help as well," a bewildered West Side alderman complained. "And what happens after it's not front-page news?"

In the midst of this blitz, Mr. T dropped by Cabrini-Green with
his signature mohawk, feather earring, and bib of gold chains. The *A-
Team* star stepped out of a Rolls-Royce, telling children that a car like

that could be theirs if they studied hard enough. It was just two weeks before the 1992 presidential election, and Bill Clinton stumped at Daley Plaza downtown, asking supporters to help him oust President Bush, who he said would never pay for the additional cops needed to keep Chicago safe. "We owe it to Dantrell Davis!" Clinton shouted to cheers.

One of the more elaborate dedications to Dantrell Davis was an architectural contest held by the *Chicago Tribune*. The newspaper put out a call for designs of a "safe, livable" Cabrini-Green. It was an ideas contest; the winner would not be hired to design anything, but the hope was that the results would offer a template for the actual remaking of the Near North Side development. "Bricks and mortar didn't kill Dantrell Davis," wrote Blair Kamin, the paper's architecture critic. But Kamin did blame the avant-garde modernism of tower-in-the-park public housing for its part in turning Cabrini-Green into an island of poverty. "Apartheid, American style," Kamin wrote, adding that these designs imposed "an ideological straitjacket on the free-flowing fabric of big-city life."

One of the three hundred entrants in the competition was a second-generation architect named Marc Amstadter, the son of Lawrence Amstadter, the original architect of the Cabrini Extension towers. The younger Amstadter said that his father's work was held up as the embodiment of an ill-conceived, faulty design. He wanted to win the contest to liberate himself—and his father—from the trauma of this past. Another entrant proposed outfitting Cabrini with an amusement park: the Big Hell Ferris Wheel. Someone else envisioned a medieval village with dramatic spires topping the high-rises. Yet another suggested connecting Cabrini to the tidal currents of Lake Michigan. But the winners all adhered to the New Urbanism that was then coming into fashion, taking their cues from old-school town planning and the antimodernism prescribed by Jane Jacobs. The winning team, a group from Fargo, North Dakota, of all places, drew up a traditional small-scale cityscape, reattached to the street grid, of low-rise rowhomes built around a neighborhood square that included commercial buildings, several schools, churches, and a daycare center. It would be an area no longer set at a remove or distinguishable from its surroundings. A half century later, it demonstrated a return to the country's first model for public housing, the Cabrini rowhouses redux.

In the past, demands to knock down Cabrini-Green and the rest of Chicago's high-rise public housing elicited charges of racism or shrugs of helplessness. Almost 100,000 poor, black residents lived in the city's public housing, with an additional 44,000 on the CHA waiting list and 70,000 homeless. Where were public housing residents supposed to go? In the most racially divided city in America, however, Dantrell Davis created a bit of consensus. "Now is the time to start tearing down the Chicago Housing Authority high-rises," declared an editorial in the *Tribune*. "Now while the memory of Dantrell Davis, shot dead at Cabrini-Green on his way to grade school, still burns our civic memory." Like Ayn Rand's protagonist in *The Fountainhead*, those who formerly supported government housing for the poor were ready to fill the towers with TNT and be done with them. In a survey conducted by the *Defender*, two-thirds of the respondents thought Cabrini-Green should be demolished. The city's newspapers opened up hotlines for callers to suggest solutions. Vince Lane looked to expand an experiment along the lakefront on the South Side in which CHA towers had been rehabbed and filled with a fifty-fifty mix of public housing and market-rate units. "If we can take Cabrini and turn it around without wholesale displacement of the people who live there, I think that can be a blueprint for this whole country," he announced. Gary Becker, the University of Chicago Nobel Prize–winning economist, suggested that Cabrini privatize, with the government selling the apartments to their current occupants. The CHA floated the idea of buying single-family homes in the suburbs and moving residents there.

Cabrini-Green did in fact need to be made safer and more livable,

and maybe even replaced entirely with something better. But acting
out of fear, spurred on by a moral panic, almost guaranteed that those
who lived there would lose out with whatever succeeded it. Looking
beyond the Dantrell Davis memorials to see the underlying problems
that caused the current conditions in the city's public housing—the
segregated site placement and overabundance of children, the mismanagement and poor maintenance, the deficits and declining for-
tunes of cities, the loss of nearby jobs and the departure of working
families, the gangs and drugs and the stigma—seeing any of that
had become nearly impossible. Fewer and fewer Americans believed
they had a collective responsibility to provide enough for those who
had too little. Most serious debate about inner cities, social services,
and housing subsidies gave way to tough talk of police, prisons, and
demolition. The mayor of suburban Bolingbrook let it be known, just
in case, that his village of forty thousand would be accepting zero
families from Cabrini-Green. The December after Dantrell's death,
word spread at Schiller Elementary that there wouldn't be enough
donated Christmas gifts for all 325 students. In their desperation,
some parents began to hoard the remaining presents. With the name
Cabrini-Green attached, the "Grinch parents" became national news,
a story of public housing dwellers so debased that they would steal
Christmas itself. One of the television news anchors reported sorrow-
fully, "Not even the season of giving could protect the needy from the
greedy at Cabrini-Green, one of the poorest and meanest of housing
projects."

SEVERAL MONTHS EARLIER, around Easter of that year, Marion
Stamps awoke from a dream. The Cabrini-Green activist looked at
the clock—4:00 a.m.—and then she began writing out the letter that
had appeared to her in her sleep. She was not a churchly person;
she didn't spend her Sundays in the pews. Her profane outbursts
could startle even the gang members she worked with. But she did
consider herself spiritual. "The Lord was revealing to me what He

wanted me to do," she said of the dream. That entailed hosting a four-day feast for the Cabrini-Green community, a unity festival, breaking bread with the "street organizations," which is how she referred to the gangs. The letter she copied out was a plea not just to them but also to everyone at Cabrini-Green. All bore some guilt for allowing the damage to their community, herself included, even if only by resigning themselves to the appalling normalcy of gun battles and the inevitability of their collateral damage. The letter addressed first the men: "You say you sell drugs because you can't get a job. You brothers will do anything but stand up and be the brave black men you were born to be." Then she called out the women: "You should be sick and tired of being sick and tired. Sisters in the ghettoes, trenches, valleys, or whatever you want to call where you live and raise your children, get raped every day for real." In a Cabrini-Green divided by gang boundaries, Stamps's Tranquility-Marksman was directly across the street from the two Cobra Stones–controlled towers and alongside white high-rises overseen by the Gangster Disciples. She insisted that her community center be a safe haven, and she regarded herself as a leader of all Cabrini-Green, as well as someone who fought for the rights of poor black people throughout Chicago and the world. She signed the letter Queen Nzinga, after the seventeenth-century Angolan monarch who defeated the Portuguese in battle.

Stamps was forty-seven in 1992 and had lived in the area for almost thirty years. That August, a week after the murder of Laquanda Edwards, she joined Cabrini residents as they marched from their home to city hall. "The residents are entitled to protection and security as United States citizens," a coalition of Cabrini-Green tenants wrote in a statement they delivered to the mayor's office. The group pointed out that guns and drugs were transported easily through the development, that police didn't patrol the buildings, and that vacant apartments and abandoned floors were never sealed off. "The citizens of Cabrini-Green have [exhausted] all options at their disposal regarding this frightening situation and have come to you for effective leadership and firm action." In September, Stamps hosted her annual Labor Day block party in front of her center, an event that she held for twenty-two consecutive years. It included games for children, burgers, ice cream, and music. Stamps talked about her letter and the need for residents to take responsibility for ending the violence. At the fields across the street, the J. P. Morgan Ewes defeated the First Chicago Near North Kikuyus to win the Cabrini-Green Little League championship. Then, a month later, Dantrell Davis was killed.

It was a tragedy that could hardly be called a surprise, and Stamps was less shocked than enraged. She was someone who normally used her tirades strategically; she fired off broadsides to disrupt meetings or attract the media or to announce that she would not be bullied, not by the Daleys or Charles Swibel or Jane Byrne or anyone. But there was nothing tactical in her fury now. She didn't know Dantrell Davis, she'd never spoken to his mother, but his killing shattered her all the same. Weeping and cursing, she made her way across Division Street, over to 500 W. Oak. She needed to see the boy's family, share in the heaviness of the loss. On her way there, she passed gang members outside their buildings, telling them she was serving notice. "Not another motherfucking baby is going to die," she barked. "Since y'all can't figure it out, I am."

"What, Marion Stamps? What?"

"Enough is a motherfucking 'nough." She sent word that she wanted to talk to the heads of the gangs. She told them to come to Tranquility.

The next week, just after Mayor Daley declared war on the criminal element at Cabrini-Green, Stamps held a press conference of her own at a little Missionary Baptist church beside the Marshall Field Garden Apartments. Daley had an eleven-point Cabrini plan; she presented a fifteen-point *peace* plan for Cabrini. Rather than boarding up the high-rise where the rifle cartridge was found, she wanted the building repurposed to house an alternative high school, a library, a youth shelter, and a drug treatment facility. She demanded that empty

apartments in other high-rises be repaired and that families who
needed homes be moved into them. She asked that the local schools
be fully staffed and the neighborhood stocked with a food co-op, a
movie theater, a bowling alley, and an arcade. Most communities took
these sorts of amenities for granted; at Cabrini-Green, their addition
could mean the difference between life and death. Stamps pressed
her first point most of all, calling on street organizations at Cabrini
to "cease all negative activities . . . to declare a citywide truce, tell
the black community you are sorry for all this pain and destruction
you have caused us." Months earlier in Los Angeles, after the riots
following the verdicts in the Rodney King beating, warring gangs had
honored a cease-fire. If the Bloods and Crips could pull it off in Watts
and South Central, then why couldn't guys from Cabrini-Green and
all over Chicago do the same? Stamps was ready to ensure that a
truce stuck: "Before I let anybody break the peace on this Near North
Side, they'll have to shoot me down like a dog in the street, 'cause I'm
comin' at 'em. I'm coming at them. You're not going to do it."

When Wallace "Gator" Bradley heard about these calls for a
cease-fire, he believed he was the one who could sell the idea on the
streets. A South Sider who seldom traveled to Cabrini-Green, Gator
knew of Marion Stamps through her activism and admired the way she
refused to back down, especially to men. "She was like a present-day
Harriet Tubman," he would say. A former enforcer with the Gangster
Disciples, Gator saw himself as a diplomat for peace as well. He'd
grown up with Larry Hoover on the South Side, and they served time
together at Stateville in the 1970s. After his release, Gator founded
a public relations firm, LeGator Productions, and he worked as an
aide to Cook County commissioner Jerry "the Iceman" Butler, the
crooner from Cabrini-Green who'd gone into politics. In 1989, Illinois governor James Thompson, a Republican, granted Bradley a
pardon, wiping clean his record, citing his work with children, and
Gator ran for alderman of the ward that included the Robert Taylor
Homes, losing in a runoff to the incumbent. He also continued to act

as a spokesperson for Hoover, who was now leading the gang from a minimum-security prison six hours south of Chicago.

In the summer of 1992, Gator was visiting the Temple Mount in Jerusalem's Old City, when he experienced an epiphany. Staring at the Wailing Wall, thinking about the divisions between Jews and Arabs, those who identified with the six-point star and those who followed the star and crescent, Gator could suddenly see how little separated the black gangs back home. Whether you were Folks or People, you spoke the same language, shared the same culture, were descended from Africans and Southerners. Upon his return to Chicago, Gator started showing guys how the five-point star and the six-point star could be combined into a ten-point star by virtue of adding a single letter, the letter "L", for Love. It was hokey, but people were killing one another over these minute distinctions. Gator needed to demonstrate that these symbols were arbitrary and mutable, that new collective identities could be formed this easily. He printed up white buttons with an image of two clasping hands surrounded by the logos of the city's major black gangs and the words "United in Peace."

In response to Marion Stamps's proclamation, Hoover decreed that there would be a truce. Shorty Freeman did the same, as did the heads of the other black gangs. In their different sections of the city, Gator and Stamps and others spread the news of the peace from street corner to street corner. Any rank-and-file member who didn't honor it would be violated by his own crew. Someone who broke the treaty and was incarcerated would have to answer for his crimes behind the prison walls. Gator told young Gangster Disciples that he didn't want them or their mothers coming to him after the fact, begging him for another chance; there would be nothing he could do for them.

At the end of October, three weeks after Dantrell Davis's killing, an actual treaty was signed at a ceremony held at Tranquility-Marksman. The leaders of each gang—every "nation"—showed up with an entourage of twenty or thirty people, filling the great hall in the back of the center. The gang chiefs brought along not only their generals but also their mothers and grandmothers, their wives, girlfriends, and children. Cousins who were in opposing gangs embraced for the first time in years. Al-Jami Mustafa, the leader of the Mickey Cobras, read from a five-page document, challenging those assembled to prove that they were descendants not of Al Capone but of great black kings. Although Stamps hosted the event, she spoke little that day. What she asked was that people think of Dantrell Davis and all the other children who'd lost their lives to the violence.

At Cabrini-Green, no shootings occurred in November. Zero in December. Two days before Christmas, a nineteen-year-old walking with his girlfriend on Larrabee was jumped by five guys and beaten with bats. The crime was notable for being the first serious assault reported in two months. For eight months, Cabrini-Green saw just one shooting and no homicides. Citywide, the year-to-year homicides dipped slightly. The truce included only the black gangs, and it didn't address intra-gang violence or incidents that had nothing to do with gangs. Police reported 570 major crimes, including two murders, at Cabrini-Green in 1993, compared with 717 major crimes and seven murders the previous year.

Wanting credit for their work, Gator and others sought a meeting with the mayor, but Daley dismissed the treaty as a ruse: "They sell drugs. That's not a gang truce. They sell weapons. They extort money from merchants. They threaten people. Forget about it." Why, then, were there fewer murders at Cabrini-Green? Simply put, Daley said, "The heat was on." The police had Cabrini on lockdown. In the *Tribune*, the columnist Mike Royko lampooned the idea that gangs should be commended, saying it had also been a long time since he'd shot anyone and there were no parades in his honor. "Gang leaders and their social activist mouthpieces apparently believe that because they haven't killed each other for a month they deserve special recognition," Royko wrote. "Most Chicagoans endure years, decades, entire lifetimes without shooting another human being. Imagine their frustration. . . . The average bungalow owner in this town never kills anybody."

Eric Davis, the police officer known as 21, said that publicly police officials couldn't praise the gangs. But privately his superiors instructed him and James Martin, Officer Eddie Murphy, to keep the gang truce in place as long as possible. Children were jumping rope and riding bicycles in Cabrini-Green playgrounds. Students were walking to stores without scanning the upper floors of buildings for snipers. Davis and Martin helped start a baseball Little League, with teams from the different CHA housing projects competing against one another. No way that could have taken place before the treaty. Years later, after he'd retired from the police force, Davis would admonish himself for squandering an opportunity to change Cabrini-Green in a more lasting way. They could have used the relative calm as a lure to bring in employers, job training, drug programs, GED classes. "Was it a good try?" 21 said. "Yeah. But so what? We didn't sustain it. Ultimately it failed."

Gator helped form a national peace movement, with the Reverend Jesse Jackson, the Nation of Islam's Louis Farrakhan, and the head of the NAACP joining him and hundreds of other current and former gang members at summits across the country. They met in Kansas City, Cleveland, DC, Minneapolis, and back in Chicago, insisting that the country couldn't arrest and imprison its way out of a crisis of youth violence. In January 1994, a day before Bill Clinton delivered his second State of the Union address, Gator Bradley and Jesse Jackson drove a white Cadillac to the White House to advise the president on the pressing issue of urban violence. It's unclear what the president took away from the meeting. Later that same year, Clinton signed into law the most sweeping crime bill in the nation's history, greatly accelerating what was already a twenty-year trend toward mass incarceration. The crime bill funded the construction of new prisons, enshrined a raft of additional mandatory minimum sentences, and eliminated many prison education programs.

The following year, federal officers in Chicago charged thirty-nine members of the Gangster Disciples with conspiracy. Larry Hoover, charged as well after agents recorded him conducting business from prison, would be transferred to the supermax facility in Florence, Colorado, where he joined Jeff Fort; the Unabomber, Ted Kaczynski; the 1993 World Trade Center bomber Ramzi Yusef; and a short list of other high-profile terrorists and organized crime bosses. All of those arrested in Chicago were reputed Disciples leaders, but with them gone the police didn't so much rid the city of the gang as eliminate its top-down authority. Chicago gangs splintered into hundreds of cliques, smaller subgroups that warred with one another. At one of the national peace summits he led, Gator blessed the convocation, praying for the "nonviolent movement" to prosper before turning to the topic of his detractors. "Father, I ask of you that all those nay-sayers, all those agents provocateurs, all those who will stand in the way of this peace, I ask that you blind them, snap the limbs in their bodies, and wipe them from the face of the earth. Amen."

IN THE MONTHS and years that followed her son's death, Annette Freeman alternated between thoughts of retaliation and of the stupefying fact that Danny was gone and she'd failed to protect him. She blamed everyone, though most of all herself. Why hadn't she listened to him the night before when he talked about his fears of dying? Why hadn't she eased up just that once and let him stay home from school? On the thirteenth of every month, she had a death wish. The only man she ever loved died September 13; then her son, the only child she ever had, died the same day a month later. She carried a gun with her, ready to shoot. She'd see a stranger and think, *You want this? Let me show you how much hate and pain I got inside me. I can crush you.*

Burt Natarus, the alderman of the area that included Cabrini-Green, had clipped many of the newspaper stories about Dantrell Davis. He sent them in a carefully arranged packet to Annette: "I write to let you know that we, in government, have not or will we ever forget your son, Dantrell." He helped pass a resolution in the city council to rename a stretch of Cleveland Avenue outside Jenner school, just feet from the shooting, Dantrell Davis Way. Annette attended the dedication ceremony, along with Jesse White and other politicians and city luminaries. She was later given a replica of the DANTRELL DAVIS WAY sign. But she stored it in a car she was living out of, and when the car was towed she lost it and her other belongings.

People told Annette they would help her with a job, an apartment, a book about Dantrell's life. She never saw any of it. And no one ever suggested counseling. No one helped cool her overheated mind. At one point she was arrested on a weapons charge and served a year in prison. She needed a gun, she felt, since no one else was out there protecting her. In prison, though, she was able to reflect on her purpose in life. She knew she could take the easy way out and become a drug addict, a killer, a prostitute. She could go and commit suicide because she'd lost everything. But she needed to represent her son in a better way. She didn't want to bring down his name. She was the mother of Dantrell Davis. She would do better.

Annette sometimes visited Danny's street sign at Cabrini. She liked that sixty years from now or in a hundred and sixty years, no matter how much the neighborhood changed, someone walking over there would look up and say, "Hey, we're on Dantrell Davis Way." She came to believe that his death had happened for a reason. That was why he had smiled at her that morning and said he was doing something. Everything led up to that moment. He was sacrificing his life for others. A citywide truce lasted for two years. Many young people who might have been killed were spared. The gangs kept it going because of Danny. And if God gave His only begotten son, who was she to ask questions? That was her comfort. She had to look at it that way. He died to save others. Dantrell Davis changed Chicago forever.

PART THREE

ROTATIONS ON THE LAND

Cabrini Mustard and Turnip Greens

DOLORES WILSON

VINCE LANE SHOWED up at Dolores Wilson's church in February 1993 to talk about the impending demolition at Cabrini-Green. He said change was coming—Dantrell Davis's death, four months earlier, guaranteed it. High-rises were going to be torn down. Two of the twenty-three towers had already been cleared of tenants and boarded up. The world of gangs and drugs had to go. Forty residents had come to hear him at Holy Family, and Lane wanted them to understand that all the attention from Dantrell's murder also meant there were tens of millions of federal dollars available to rebuild. Lane asked the tenants to partner with him in making the *new* Cabrini-Green. He promised that nothing would be done without their input and approval.

To give a sense of what a revamped Cabrini might look like, Lane pointed to what he'd done at Lake Parc Place, a public housing development on the South Side. Four of the six high-rises along the lakefront were vacated and eventually demolished. In the two rehabbed towers, only half of the apartments were designated as public housing; the rest were rented at slightly subsidized rates to working-class and middle-class families. When the two fifteen-story buildings reopened, in 1991, all 282 apartments were filled, with public housing recipients and better-off renters living next to each other on every floor. The rehab resulted in a net loss of more than 550 units of public housing. But Lake Parc Place was now livable, Lane explained. You had people of varying incomes who *wanted* to live there. "High-rises weren't the problem at Cabrini-Green," he said. "Rich people all around them lived in high-rise apartment buildings. The problem was the high concentration of poverty."

Lane was a visionary. He could see where public housing policy in the country was headed. Only a couple of years earlier, Jack Kemp, the head of HUD under George H. W. Bush, said he refused to be remembered as the "secretary of demolition." Kemp had founded the federal HOPE initiative not to tear down high-rises but to hand over responsibility for the properties to the people who lived there. HOPE stood originally for Homeownership and Opportunity for People Everywhere. But over the next decades, HUD would award cities tens of billions of dollars in HOPE VI grants to knock down their public housing high-rises and replace them with mixed-income developments like Lake Parc Place. Demolition became something that local and federal politicians embraced proudly.

While public housing was "never more than half alive," as Catherine Bauer called it, the aversion to entitlement programs would come to root ever more deeply in the American mainstream. After winning a majority of seats in Congress in 1994, Republicans said they planned to scrap the entire Department of Housing and Urban Development. President Clinton, touting his cuts to the welfare system and proclaiming that "the era of big government is over," preemptively reorganized HUD. The agency promised to "infuse market discipline" into public housing. A new law mandated that any large public housing development with a vacancy rate of over 10 percent undergo a test to determine whether a rehab made sense economically. Of the 100,000 units across the country that failed, a sixth of them were in Chicago, a total of 18,500 apartments, twice as many as in any other city. At Cabrini, only the rowhouses passed. Congress then voted to end the requirement that cities add a unit of public housing for every one torn down, ushering in the era of widespread demolition. By 1999, HUD would boast that HOPE VI had eliminated 50,000 units of public housing nationwide; a decade later, the number had doubled. Every city got in on the action: Philadelphia, Atlanta, Baltimore, Newark. But none would knock down as many as Chicago.

Mixed-income developments like Lake Parc Place, with their blended populations and low densities, could house only a small percentage of those who had lived in the demolished public housing high-rises. But these new developments would also serve the purpose of clearing the slums—replacing the same public housing buildings that were used to clear the slums of a half century before. The majority of residents leaving public housing were given vouchers to find apartments on their own. The Section 8 program allowed cities to move away from building and operating actual public housing complexes; the government would instead pay private landlords to rent to qualifying families. By 2000, the CHA was calling itself a "facilitator of housing opportunities," and its holdings would include 48,000 vouchers and half as many actual public housing units. The rents the government paid to Section 8 landlords in Chicago were based on an average of the city as a whole—meaning too little for a home in a diverse neighborhood with strong schools and low crime, but, it turned out, somewhat more than the asking price of a place in an overwhelmingly black and poor neighborhood. Landlords in these areas saw a profit in a guaranteed government check, and so the solution for high-rise public housing would also re-create some of its same mistakes. Many of the families relocated with vouchers ended up in areas almost as uniformly black and poor as the government housing they left.

Vince Lane couldn't know how vehemently Cabrini-Green residents would fight to save their tragic home over the next quarter century, or who among them would get to stay on the land in new mixed-income buildings, or what would happen to those who were dispersed. But he knew that the name Cabrini-Green was a catalyst for this change to come. He announced as he prepared the $50 million grant application for the first phase of the development's overhaul, "Cabrini symbolizes all that is wrong with public housing."

The residents of 1230 N. Burling who listened to Lane were unmoved by his mixed-income vision for Cabrini-Green. Their building had the same high concentration of poor and black families as the other towers, and yet they were already partners in its restoration. HUD had awarded the 1230 North Burling Resident Management Corporation $6.8 million to oversee a full-scale renovation of the high-rise. They were getting a new roof, new pipes and trash chutes, a new heating system. They were replacing all the windows and rehabbing every apartment. And like anyone renovating a home, Dolores Wilson and the other managers obsessed over the details. They selected colors and styles, changing their minds ten times before settling on their picks. Blinds: white Levolor. Tile: mauve vinyl. Elevator entrance: terra-cotta. Exterior facade: eggshell. They interviewed engineers, contractors, and suppliers. They devised a five-phase renovation schedule, with families slated to move temporarily into rehabbed vacant apartments in the building while their own units were completed.

Bertha Gilkey warned them that they needed to document how every cent was spent, since people were going to assume that they were incompetent or, worse, thieves. If the renovation was over budget or shoddily done, the residents were sure to be blamed. "I was scared," Dolores would say. "I figured we were going to mess up on this or that." So it was a milestone worthy of jubilation when their resident management corporation finalized its very first construction contract, in 1992, for the installation of the weather-sealing double-pane windows and the repair of the tower's exterior concrete. Dolores wrote in the flyer announcing the celebration, "This ceremony would symbolize the success of the RMC's hard work, cooperative spirit and participation in the comprehensive modernization program." As further proof of how much they'd accomplished, she included in the invite the contract number for the work and its dollar amount, $1,274,000.

The general contractor that Dolores helped pick came highly recommended. Shah Engineering had numerous jobs with the city and state. When the work progressed slower than scheduled, the firm's owner, Manu Shah, assured her that there were always unforeseen delays. He said his crew had discovered lead-based paint in many of the apartments. The craftspeople he hired complained as well about cockroaches, saying they weren't paid to be exterminators. When Shah bid on the project, he and his subcontractors had to detail how they would hire residents. People from the building wanted to work. Dolores and her team readied them, sending residents to training sessions, giving them tips on how to be interviewed, and submitting their names to the CHA and HUD. But the contractors and subcontractors came up with excuses for why they didn't hire more people from the lists provided. One morning when the painters and drywallers showed up for work, a group of residents barred their entrance into the building. The tenants were shutting down the worksite until the contract was honored. Days passed, and then weeks. The CHA assigned an official to mediate the dispute. Shah agreed to another round of interviews. One subcontractor hired six additional tenants. The roofer took on two residents at $13.57 an hour. Work on the building resumed.

A few years later, after a long federal investigation, Manu Shah pleaded guilty to bilking the city and state out of $10 million. His engineering firm had been awarded more than sixty government contracts with Chicago alone, and it was caught overcharging on the jobs. The residents of 1230 N. Burling could only shake their heads. "That's Chicago," they'd say. They'd seen it before.

But as far as the renovation of their building went, Dolores and her neighbors were thrilled with the results. Amid the public outcry to tear down Cabrini-Green, their high-rise looked new. The outside walls were restored to the color of creamy vanilla. The mechanicals were updated. And Dolores loved her new kitchen and weather-sealing double-pane windows. "Everyone wanted to move into the Burling," she'd say. The building didn't just look good. For the residents, it was a monument to what could be accomplished when they were given the power to shape their own lives. "I wish Oprah had said 'There are no children *here*,'" Dolores would pronounce. The Alex Kotlowitz bestseller about the Horner Homes had been made into a television movie starring Oprah Winfrey as the mother of Lafeyette and Pharoah Rivers. "We would have been all over her," Dolores went on. "Children still live here."

IN THE FIRST issue of *Voices of Cabrini*, in April 1993, the newspaper's co-editor, Mark Pratt, welcomed readers by expounding on the publication's name. "How often have you sat and either watched a report on television or read an article in a newspaper about Cabrini-Green and said to yourself, 'I wish they had spoken to me,' because you felt you had something to contribute or debate?" Six months had passed since Dantrell Davis's murder, and Pratt was sick of the police sweeps and his neighborhood being called out as a paragon of inner-city horrors. He'd been a small child in 1972 when his family moved to the rowhouses, and he was ready to set the record straight on many things. He wrote of the homegrown gang truce, "It is a well acknowledged fact by some police officers within the Cabrini community that the peace treaty is really what has changed the safety of Cabrini and not the storm troopers, or Nazi-like tactics of the lockdown." One contributor, addressing the imbalance in the media coverage of his home, wrote, "When something bad does happen in Cabrini-Green, and you do a story on it, please leave Dantrell Davis out of it. LET HIM REST IN PEACE."

Every three or four or five weeks, whenever an edition was completed, *Voices of Cabrini* would be filled with lists of graduates from the local public schools, tutoring programs, and GED classes, updates on the Little League season, and congratulations to tenants like Natalie Howard, of 911 N. Hudson, who took first place in a playground safety contest. It included poems, movie reviews, and remembrances of those who'd passed. The paper shared tips on how to identify child abuse and how to deal with an infant who wouldn't sleep. Sidebars explained tenants' rights and provided information on the area's churches and legal-aid clinics. A lifetime Cabrini tenant, reflecting on his journey into and out of addiction, thanked local role models, notably Hubert Wilson, whose drum and bugle corps had inspired him to get clean and to teach drums at Sojourner Truth Elementary. "A family man," he wrote in memory of Dolores's husband, "who still took time to be an outside father to boys who didn't have a father."

The newspaper grew out of an unexpected alliance between Cabrini-Green and the wealthy North Shore suburb of Winnetka. Peter Benkendorf, who operated a Chicago nonprofit focused on civic engagement, made the match. As he followed the news of Dantrell Davis, he recalled a shooting that had occurred in Winnetka four years earlier. A thirty-year-old woman named Laurie Dann, after trying to poison dozens with arsenic, walked into an elementary school with three guns, shooting six and killing one child before taking her own life. The events led to a book—*Murder of Innocence*—and a made-for-television movie starring Valerie Bertinelli as the unhinged killer. "School shooting" wasn't yet a thing, but Benkendorf imagined that Cabrini-Green and Winnetka shared an unexplored bond. He connected with Henrietta Thompson, a Cabrini tenant who'd assisted the *Candyman* producers when they filmed on location at the development. They drove out to Winnetka and met with a group of women who'd started a community newspaper following the Laurie Dann incident as a form of therapy and a way to rebuild trust. At Cabrini, no one received grief counseling, not even Dantrell's young classmates, and residents were being told there was little there worth saving. Thompson liked the idea of the newspaper.

Thompson and Pratt shared the editing responsibilities. Another ten Cabrini residents joined the staff, and during the first year of publication nearly a hundred people from Cabrini-Green contributed. They laid out the pages at a nearby ad agency where Benkendorf had worked, printing a thousand copies of each issue. Pratt borrowed a used car to deliver stacks of the papers to the high-rises and row-houses, to local churches and community centers.

Jimmy Williams, who as a teenager had been the drummer in the Electric Force Band, wrote a regular advice column, often of the tough-love variety: "Peer Pressure? What the hell is that? If you feel so low about yourself that you must follow the Joneses, lay down and die." Godfrey Bey, who'd grown up in public housing on the West Side and owned a fast-food seafood restaurant, wrote a cooking column called "Come and Get It!" He presented recipes for such dishes as spaghetti conquistador and Cabrini mustard and turnip greens.

The reviewer who covered the music scene at Cabrini-Green— "Chi-town's capital hip-hop headquarters"—mailed in his pieces from rural Illinois, four hours southwest of Chicago. K-So, who'd written lyrics for the Slick Boys, was in a state prison serving five years on a drug conviction. He managed to keep up with the trends (he rated Naughty by Nature's "Hip Hop Hooray" the "dopest single of 1993"), and what he couldn't verify from the penitentiary he made up. Born Pete Keller in a section of the Near North Side neighboring Cabrini, he renamed himself "K-So" at twelve, for "Knowledge, Strength, Opportunity." As a teenager, he decided to make Cabrini-Green his home, residing over the years in the rowhouses and twenty-one of the twenty-three high-rises. He loved the public housing project. He hated it, too, since many of the people he knew and cared about there ended up dead, in prison, or strung out, sometimes on the drugs he sold them. "Cabrini-Green has layers," he'd say. "Cabrini has so many fucked-up multiple layers."

In 1995, when Keller was paroled from prison, he hopped a Grey-hound back to Chicago and finished out his sentence under house arrest in a Cabrini-Green high-rise. *Voices of Cabrini* had stalled by then, and K-So restarted the paper with the zeal of someone who'd thought about little else for thirty months in an eight-by-ten-foot cell. In one of the many articles he wrote, he tried to make sense of a visit to Cabrini-Green by US congressman Christopher Shays, a Connecticut Republican who showed up at the housing complex unannounced over a Labor Day weekend.

Ever since Jane Byrne's three-week residency, numerous politicians had made a point of coming to Cabrini, usually to add grist to whatever housing or welfare reforms they were pushing. At 1230 N. Burling, Dolores Wilson regularly toured congressmen and senators. She met Jack Kemp so many times, she said, that she was willing to switch over to the Republicans if it meant sending him to the White House. "I don't care what Jesse Jackson or anyone else says," Dolores would say. "I'm voting for Kemp because I know what he's for."

Keller was happy to play host to Shays. Fair skinned and straight haired, K-So was raised by adoptive parents who were white, and at various times he lived as a white person, a black person, and a Latino. For these reasons, he thought of himself as a bridge between cultures—and between Cabrini-Green and the outside world. He walked the congressman around the land at Cabrini, introducing him to residents. When Shays seemed eager to see more, K-So asked if he wanted to stay the night. Shays did. He slept on the couch in K-So's living room. "'Honey, it's incredible. I'll tell you about it when I get home' whispered Christopher, as I overheard him calling home to his wife from an old pay phone at the bottom of my building," Keller wrote. The two men spoke late into the night. Shays asked about living conditions and what sorts of jobs people had and what the worst thing was that Keller had seen. Shays was a Republican who denounced federal control of public housing, a New York University MBA, a Christian Scientist who was still married to his childhood sweetheart. The guy seemed as foreign and inscrutable to Keller as the people at Cabrini must have come across to him. "I wondered what he might have been thinking about us, his stay, future concerns or if he just wanted to say that he'd spent the night in Cabrini-Green," Keller mused.

The next morning Shays thanked K-So for the hospitality. The couch he found comfortable, but he did have to ask about the thick fug of heat in the apartment, the lack of any circulating air. It was unbearable. "How do you guys manage?" The question seemed like a bit of life imitating the art of *Good Times*. In an episode of the sitcom, a white official from the housing authority calls on the Evans family and is shocked to find the elevator out of order—he can't catch his breath after hiking the seventeen flights of stairs. He learns, too, that the heat isn't working, the refrigerator is broken, and no water is flowing from the taps. "I hate to bring you people down," he says to the amused family, "but it's awfully depressing to hear things like that." Then word reaches the Evans apartment that two gangs are fighting outside the towers, forcing the official to stay for dinner. Outraged to discover that nice people are suffering in these disgraceful conditions, he announces that he is going to bring changes. He's a top administrator at the district office, and he will make it happen. "You have my word, everything that's wrong in this building is going to be fixed." When the gang fight ends, he hurries out the door, promising over his shoulder that the Evans family should start to see repairs . . . in thirteen or fourteen months.

Keller didn't reply to Shays's question, or at least he said he didn't in the article, which he ends with an oddly revealing combination of conviction and doubt, a silencing of his Cabrini voice. He writes, "I wanted to comment, but I knew inside that he too knew the answer. Didn't he?"

VINCE LANE RETURNED to Cabrini-Green three weeks after speaking at Holy Family. On the campus of the nearby Moody Bible Institute, Cabrini residents joined him at an all-day "planning summit." Asked to enumerate their concerns, tenants said they worried that they were being tricked off their land. They were allowed to live on the Near North Side so long as the area was full of crime and poverty. Now with investment and upkeep, with improved policing and new amenities, they didn't believe that they would be allowed to stay to enjoy it. Lane led three more meetings at Cabrini the following month and a town hall for all residents.

Did the tenants feel like they were heard? They did. They joined committees and neighborhood advisory councils. They met with an independent facilitator Lane hired to keep the dialogue going. They were convinced that they were included in what the CHA was now calling the "rebirth of Cabrini-Green through resident empowerment partnership." This first phase of Cabrini-Green's development would cover only nine of the seventy acres, just the northern end of the Cabrini Extension site along Division Street. For more than a year, the residents in the "empowerment partnership" dutifully completed the application for the $50 million HOPE VI grant. In the days before turning in the application, they traveled by bus to a suburban Hyatt hotel, thirty minutes outside Chicago. The weekend planning session began with a motivational screening of *Fired Up!*, a documentary that had been made about the 1230 N. Burling resident managers. Then in groups they hashed out the particulars of how security would operate in the redeveloped section of Cabrini, how the construction could be used to hire residents and support resident-owned businesses. They discussed the ideal blend of public housing and market-rate units in the mixed-income replacement buildings. Tenants wanted the rebuild to include the maximum number of public housing units, and the city argued that there was a threshold at which the neighborhood could lose the stability and diversity it had recently gained. The city hoped the mix in the new buildings would hew closer to a Realtor rule of thumb that white people wouldn't move into any neighborhood that was more than 30 percent black. The residents wanted guarantees that they could stay in the area, even during construction; the city insisted that it had to support the private developers who would build the mixed-income housing and take on the risk of selling or renting out a majority of the units.

In the end, everyone agreed to the demolition of three towers. In the new buildings replacing them, 40 percent of the units would be reserved for public housing families. The proposal also earmarked funds for a battery of new and improved social services, for a resident-owned security company, and for job training and other resident businesses. Mayor Daley sent the group a letter in support of their application, and one of the Cabrini leaders called it "a marvelous plan. It will better the community and give us pride."

Vince Lane didn't have the opportunity to see this marvelous plan come into being. In May 1995, HUD seized control of the CHA, forcing out Lane as part of a federal takeover of the agency. The CHA had long been a stalwart of dysfunction, remaining on HUD's troubled housing authority list since the designation began in 1978. In its most recent evaluation, the CHA managed a score of just fifty out of a hundred. (New York City, with 180,000 units of public housing, scored in the nineties.) HUD saw that Lane had spent $74 million on sweeps and other security measures over the past year, and still CHA residents were twice as likely as other Chicagoans to be victims of a serious crime. Employees of the agency were caught in various schemes involving ghost workers, falsified overtime records, and over-charging for supplies.

Lane believed he was targeted because Mayor Daley saw him as a political threat. A few years after being ousted from the agency, he was convicted for making false statements on a loan application tied to the development of a South Side shopping center. For the violation, Lane's lawyers said he deserved at most probation, since no one suffered a financial loss, but Lane received a two-and-a-half-year prison sentence. For many, the punishment seemed excessive. Lane certainly thought so: "Daley killed me off, period."

Cabrini residents adapted, joining with others to evaluate the developers bidding on the redevelopment of the nine-acre site. In the meantime, the city went ahead and tore down 1117–1119 N. Cleveland—the home of the Castle Crew, the building Brother Bill frequented, and where Dolores Wilson and her family had lived when they came to Cabrini in 1956. The building had stood for thirty-nine years, and it was the first Cabrini-Green high-rise to be demolished.

Its demise was cause for both celebration and reflection, hope and concern.

Then, suddenly, Cabrini residents who'd been involved in the redevelopment partnership for two years were shut out of meetings. They were told they'd be consulted in due time. Months passed. In June 1996, Mayor Daley made a public announcement about a revised development plan for the neighborhood that differed significantly from the previous one. Flanked by politicians and planners but notably no Cabrini-Green residents, he said eight Cabrini towers would now be demolished, not three. The rebuild would be much larger, spreading to other nearby plots of land, and would add 2,300 units of housing, in townhomes and three-flats, restoring the street grid and including new schools, a police station, a shopping center, and an expanded park. Just 15 percent of the units under Daley's proposal would be available to very-low-income families, representing a net loss of nearly a thousand public housing homes. The plan seemed less like the redevelopment of the Cabrini Extension towers than a public-private push to accelerate gentrification in an area abutting the city center. "How can they take our roots from us without our input?" Dolores's daughter Cheryl asked. "They're planning to move the poor and destitute and build for rich folks."

Five years earlier, in 1991, residents at the Henry Horner Homes had sued the CHA and HUD, accusing them of "de facto demolition"—allowing the vacancy rate to reach 50 percent and conditions to degrade so much that there was nothing left to do but tear down the buildings. In a 1995 settlement, the city was forced into a consent decree that dictated the terms of Horner's redevelopment. At Cabrini, tenants had trusted that their city partners would hold up their side of an agreement. But at Horner, the city was bound by a court order. More than half of the units were reserved for very-low-income families. Demolition and construction were staggered, and residents never had to cope with an involuntary relocation.

So in 1996, Cabrini tenants sued as well. Twenty-two charges in all, accusing the Daley administration and the CHA of violating the 1968 Fair Housing Act, which made it illegal to use federal money to maintain racial segregation, the Civil Rights Act of 1964, the US Housing Act, the Community Development Block Grant Agreement, and the HOPE VI statute. To the surprise of city government, a US district judge listened to testimony and in January 1997 ruled against the CHA's motion to dismiss the lawsuit. Eighteen of the twenty-two counts could go forward, and he issued an injunction halting further demolition at Cabrini-Green until the case was heard. It was a victory for tenants. "There is a God up above," Cora Moore said.

DOLORES WILSON

BEFORE THE FILING of the lawsuit and even before the rehab of 1230 N. Burling was finished, Dolores Wilson decided to step down as president of her building's management corporation. She was sixty-four in 1993 and tired. She'd been helping run her high-rise for more than a decade, and since Hubert's death her calendar had been filled from one year to the next. When she announced her resignation, Bertha Gilkey persuaded her to stick around a little longer. "Just another month," Gilkey urged, long enough to see through the repairs of the building's roof. Then a month passed, and Gilkey asked her to wait until other construction contracts were signed. Then she needed Dolores to travel to Washington with her to speak to lawmakers about resident management. Dolores knew she had trouble saying no. Her pastor would ask her to go somewhere, and what could she do? She'd go. Finally, Dolores was allowed to hand off the presidency to one of her neighbors.

She left the water department around the same time. The city was offering early retirement, and after twenty-seven years at the job, Dolores took it. The head of microscopy was from Bulgaria, and to Dolores she sounded just like Zsa Zsa Gabor. "This is your day, Dolores. Stop filing papers. Come here," she said. She led Dolores to a long table at the department covered with food, and people from all the laboratories came by to wish her farewell. That was a week or two before Michael Jordan's father was killed. For years to come, that's how Dolores would place the date, because she sent the Bulls star a sympathy card, and he sent her a thank-you in return. She'd gathered all the children in the Burling building to write condolences as well. She told them not to beg Jordan for anything, just to write that they were very sorry to hear about his daddy. But one of them asked for a bike. That boy used to break the wings off pigeons, and Dolores told his stepfather that children who were allowed to torture little animals would get the feeling of death in them. Eventually, it wouldn't mean anything to the child to cripple or kill a person. That's how Jeffrey Dahmer, the Milwaukee cannibal, got started, she explained. But the boy later became a preacher. "Good things can come out of bad, I guess," Dolores said. "I think at least I saved some animal lives."

Dolores had a harder time recalling exactly when her own Michael, her second child, died. "Some things you just can't . . ." She tried not to think about his death, but guessed it happened two years earlier, maybe in 1991. Mike was almost forty then and divorced. He had four children and was also raising his girlfriend's two as his own. He was with the girlfriend on a summer night, at a Chicago Avenue sandwich shop by the rowhouses. He went to buy cigarettes at a gas station on Orleans, and when he returned a guy who used to date his girlfriend was threatening her, saying he'd bash her head in with a bottle. "One thing about my sons," Dolores would say, with a mix of pride and sorrow, "they protect their women. They're gallant." The two men came to blows, and Mike ended up on top. That's when one of the ex-boyfriend's associates pulled out a gun and shot Michael in the back, close range, with a hollow-point bullet.

It seemed utterly unfair to Dolores. She had dedicated much of her life to the community and the neighborhood children. She'd received service awards from her church and the YMCA. The district police had honored her and Sugar Ray Dinke on the same day. Jack

Kemp had personally named her the HUD "Resident of the Year." All that, and look what happened to her own baby. Michael's children were left fatherless. In her grief, Dolores wanted to be left alone, but the visitors kept coming. She said little in front of them, even joked, but then she'd close herself in the bathroom and scream. Someone had to explain to her what a hollow-point bullet was, how it expanded as it traveled inside the body. Hearing that made her think differently about people.

One of the members of her church had a twenty-five-year-old daughter who'd gone missing for three months a couple of years earlier. The police refused to look for her, saying, "You sure she's not laid up with someone?" The daughter had been murdered it turned out, her body left to decay in a sewer two blocks away. The woman had a son as well who'd been shot in the back behind one of the Cabrini high-rises. That mother said to Dolores, "Ms. Wilson, you always told me to keep the faith. Now you have to keep the faith." The pastor from Holy Family counseled Dolores, saying she was the most forgiving person he'd ever known.

At the funeral service, Dolores stood in front of the congregation and said she didn't want any retaliation. She asked that Michael's friends and other folks from the building pass along that message. She attended the trial, and when the ex-boyfriend was found not guilty she hugged and kissed his mother, telling her she knew he didn't do the killing. She felt less compassion for the shooter. He was found guilty of second-degree murder and sentenced to seven years. Dolores's youngest son, Kenny, had broken into a car trunk once and stolen a few tools and received the same sentence. "There's no justice," she would say. "A gangbanger got seven years for killing my son, and Kenny got seven years for stealing a damn screwdriver and wrench."

Dolores had been hosting family reunions in Lincoln Park every August for a decade, but she stopped after Michael was murdered. Sometimes she wished her children had moved away, had left Chicago and stayed in Indiana or another place. But she didn't blame Cabrini-Green. She didn't imagine her children would have lived a better life elsewhere. A reporter who interviewed her three days after Michael's funeral asked if there was anything she wanted to convey to outsiders about her home. Dolores paused to think. She guessed she did have a message: "Tell them that there's more love over here than terrorizing."

If Not Here . . . Where?

KELVIN CANNON

KELVIN CANNON GOT off the elevator one morning, on his way out of the building, and Cora Moore blocked his path. When he was a boy, living next door in 714 W. Division, Moore often scolded him for fighting with the children in her high-rise. He was now twenty-six, and she still reprimanded him—she was in charge of resident security at 1230 N. Burling, and he called it for the Gangster Disciples there.

"You should get on my team," Cora began that morning in 1989. "Do you want to be out here and be free and raise your family? Then you have to leave gangbanging alone." She'd been delivering the same pitch for a couple of years, trying to recruit him away from the gang. "What I'm offering they can't offer you. I'm offering a chance at a better way of life." Cannon lingered in the lobby longer than usual. He'd been wondering lately whether there were other options available to someone like him, thinking that probably there were none. He'd managed to avoid a return to the penitentiary, but he was tired of the police raids, of going back and forth to jail for weeks at a time.

Cora, sensing the opening, told Kelvin that he was smarter than most. She said the building's resident management was getting started and he could be one of its leaders. He could work right there in the high-rise. Like Bo John had made him believe when he was thirteen, Moore said he could be part of something bigger than himself. "If you really want to change, help me fix the building," she told him. "Let me show you the other side of life."

"Okay," he said. "Show me."

Cannon went to the leaders of the Disciples and said he was re-signing his position. Moore talked to the commander at the police precinct and said Cannon was now working under her. She signed him up for a security guard class. In prison, he had read the Bible—first the New Testament and then the Old—and he continued to school himself after his release, looking at a dictionary in his apartment, working on his vocabulary and spelling. But he hadn't been in a classroom since he was kicked out of Cooley High. He completed the security guard program, and in his uniform and cap he monitored the stairwells and ramps of his high-rise. Some thought it hypocritical that after all Cannon had done he was telling eight-year-olds to stop fooling around on the elevators and directing young men to move from in front of the building. But he said he'd always been in charge of people at Cabrini-Green and giving orders in one form or another. He was only doing more of the same. "You turned police?" a guy from the building asked him. "Nah, my brother. I'm trying a new way of life."

He was one of the few men to go through the resident manage-ment classes with Bertha Gilkey. On the weekend retreats, he sang "We Shall Overcome" with Dolores Wilson and everyone else, and he studied budgeting and federal housing regulations. He went along on the trips to see other tenant-managed housing complexes in Boston, DC, and St. Louis. Like Dolores, he came to know politicians person-ally. "I haven't forgotten my pledge to be your 'bridge' to HUD and the City of Chicago," the secretary of housing wrote in a note, asking about the well-being of Cannon's wife and signing it, "old #15, Jack Kemp, your friend." Moore used her connections to get Cannon into a training course for CHA janitors. He apprenticed at Cabrini-Green, shadowing the electricians, carpenters, glaziers, and locksmiths. He did masonry and bricklaying, whatever the buildings needed. For two years, he took night classes at a trade school. After graduating, he started off at $11 an hour, but when he was put in charge of the janitors in 1230 N. Burling, as a union worker, he earned $17,000 a year, plus benefits. That was good money, in all senses. He earned it as a day in, day out laboring professional, on call twenty-four hours a day. He barely left the high-rise, and he'd become more like Hubert Wilson, Dolores's husband, than his old self. He joined a church and trained to be a deacon.

Cannon took pride in his high-rise's improvements. "It looked better than any other building in Chicago that was public housing," he'd brag. "You know how in that Ajax commercial, they go past certain houses that are all grimy and filthy. Then they come to the one real clean house—it's spotless. The music stops and a woman says, 'Ajax was here.' That was our building—like Ajax had been there."

J. R. FLEMING

WHEN J. R. was arrested for selling drugs on the far North Side, it was Jesse White who picked up the phone, telling the prosecutor that he vouched for the young man. "Jesse White took me under his wing," J. R. said. As part of the Young Democrats of Cook County, J. R. helped out whenever the ward boss, George Dunne, hosted a community event or when White needed guys to pass out turkeys or school supplies. In the procession at the annual Bud Billiken Parade across the South Side, he carried the banner of the ward organization. And come election time, he hung campaign posters, tore down those of the opposition, and ushered Cabrini residents to the polls. "I was being raised politically," J. R. explained. One time, the pastor at Holy Family asked J. R. if he could get his bosses to fix the sidewalk in front of the church. He brought it up, and the sidewalk was repaved. "It was an awakening," J. R. recalled. "Politics is a means to an end. My power got that done. It felt great."

Men ran the ward. Women, for the most part, ran Cabrini-Green. Marion Stamps cornered J. R. one day at a neighborhood youth center. "Don't be a puppet's puppet," she warned him. "There's always a master pulling both of your strings." She ran again for alderman in 1995, reminding voters that she'd orchestrated the gang truce and railing against the plans to tear down sections of Cabrini-Green. She called it a tragedy that "revitalization" meant moving in white and wealthier residents and moving out the neighborhood's longtime inhabitants. "It's no accident that they pieced together this ward with little pockets of black people that they intend to make disappear. It's all about displacement and reclaiming of the land," she said. "If they come at Cabrini in the daytime, they're going to come after the rest of them at night. And that's not only public housing; that's any housing that's available for poor people."

Her chief opponent in the race was J. R.'s immediate supervisor in the ward office, the chairman of the Young Democrats, Walter Burnett, who as a young man in the Cabrini rowhouses was an accomplice in a bank robbery and served some jail time. After his release, Burnett went to work for the ward organization and was mentored by Jesse White before rising through the local patronage system. He had the power brokers on his side; J. R. was one soldier in a small army of campaign workers they commanded. Burnett won easily, and Stamps decided to leave the Cabrini-Green area and move back to her hometown of Jackson, Mississippi. Her father was still living there, and she saw a fitting symmetry in attaining political office where her activism began. But at the age of fifty-one, before she'd packed a single bag, Stamps died of a heart attack in her sleep. "Stress killed her," her daughter Guana said.

About a year after Burnett's election, J. R. decided to get out of politics altogether. He had worked what he called slave hours for six straight days setting up for the 1996 Democratic National Convention. By pulling off the event without a hitch, the second Mayor Daley would simultaneously match his father as political kingmaker, helping launch Bill Clinton's reelection campaign, and surpass him by avoiding the chaos that ensued on the streets and in the convention hall a generation before. Twenty-eight years earlier, when Connecticut senator Abraham Ribicoff used his nominating speech to denounce the "Gestapo tactics" of the Chicago police then beating up protestors on Michigan Avenue, people observing the first Mayor Daley said he yelled, "Fuck you, you Jew son of a bitch! You lousy motherfucker! Go home!" (Daley Sr. said he never in his life used such language. "Faker" was what his supporters said he shouted.)

In 1968, Chicago had used tall fencing to mask the bleakness of the old stockyard neighborhood surrounding the convention center. The 1996 convention took place in the United Center, the new home of the Bulls and Blackhawks, completed just west of the Loop two years earlier, at a cost of $175 million. Daley's vision for the city was inspired by his trips to Paris: Chicago as a postindustrial urban gem, a global city, and a central tourist attraction. Here was an opportunity to showcase the revitalized Chicago he was creating, a city once again on the make. Several of the Horner high-rises across from the United Center were demolished. The city covered nearby vacant lots with wood chips and newly planted bushes. Owners of parking lots and other properties in the vicinity were pressured to put up wrought iron fencing and planter boxes. Roads were repaved and lined with antique-style streetlamps and hanging flower baskets. Bridges were painted, the exteriors of nearby schools restored, street signs added, and the train running west from the Loop over Lake Street reopened. And like in Paris, the office buildings in the Loop were lit up at night.

When Bill Clinton first ran for president, he presented himself as a "New Democrat." He derided the longtime failures of government and pledged to "end welfare as we know it," explaining that William Julius Wilson's writing on "concentration effects" "made me see the problems of race and poverty and the inner city in a different light." Running again as a moderate in 1996, angling to win back Congressional seats lost two years earlier, Clinton touted the welfare reform bill that he'd signed into law only a week before his arrival in Chicago. The federal entitlement now came with work requirements

and a lifetime limit of five years for any recipient. He sold himself as
tough on crime as well—he was putting 100,000 new police officers
on the streets and had made "three strikes and you're out" lifetime
sentences for habitual offenders the law of the land. The Republican presidential candidate, Bob Dole, referred to public housing on
the campaign trail as "one of the last bastions of socialism in the
world," arguing for its privatization and erasure. Clinton agreed that
the program needed to be curtailed, accepting that the demand-side
marketplace probably couldn't do worse than the likes of the CHA.

Jesse Jackson spoke at the convention—reluctantly, he claimed—
saying that Clinton's cuts to the social safety net were a devastating
blow to the have-nots. He said new public housing complexes were
in fact being built and maintained across the land—they were the
federal and state prisons that were home disproportionately to black
men. Jackson argued that the Democrats still needed to be the party
that fought for the social well-being of all Americans. The idea that
there were enough decent paying jobs out there for anyone with
initiative to seize was a fantasy. Citing Martin Luther King Jr., he
said there were "mountaintops" of prosperity in the center of his city.
But much of the rest of Chicago lay in a canyon. "Once Campbell's
Soup was in this canyon. Sears was there, and Zenith, Sunbeam,
the stockyards. There were jobs and industry where now there is a
canyon of welfare and despair." The national economy under Clinton
had rebounded, and a seemingly bottomless deficit had been filled.
But inequality was deepening and widening. "What is our obligation
to the people in the canyon?" Jackson demanded. "That was Roosevelt's dream, and Dr. King's."

J. R. never set foot in the United Center during the convention.
He spent his days setting up events in downtown hotel rooms where
visiting delegates gathered. His hands were calloused from hanging
posters, his thumbs bruised from tying balloons. He didn't hear a
word Clinton or Jackson said. But J. R. reflexively dismissed the Reverend Jackson as a poverty pimp, looking to advance his own career

on the plight of the poor. And as for Clinton, J. R. knew that the president hadn't stood up for Midnight Basketball, the league he played in at Cabrini-Green. Republicans attacked the puny line item in the president's $33 billion crime bill, ridiculing the idea that any tax dollars might be spent on what they demeaned as games for criminals and crackheads. Clinton scrapped the program rather than defend it. It didn't matter that the league had been started under President George H. W. Bush, one of his "thousand points of light," as a way to keep public housing residents engaged in positive activities, or that it was proven to reduce crime. J. R. was done with them all. He told Brother Jim about his decision to quit politics. "I knew you'd figure that one out," Jim said.

What J. R. focused on instead was perfecting his peddling business. He bought a red GMC Vandura, and later, when his profits allowed it, a blue one, too. Most mornings he parked one of the vans outside a Cabrini-Green high-rise, laid on the horn, and then began his rounds. "Tapes, CDs, movies!" he cried. "Tapes, CDs, movies!" J. R. shouldered a military duffel the size of a rowboat. Inside he placed cardboard runners to partition his wares, like in a case of wine: old soul and hip-hop in separate compartments, pirated video games in others. From the strap, he dangled plastic bags filled with Bulls T-shirts and caps, with socks and towels. Michael Jordan had given up on baseball, thankfully, and the Bulls were again winning titles—God's gift to the city and to its street sellers. J. R. was a one-man Chicago Bulls emporium. His wasn't some junkman's operation, either, a strung-out barker pushing a stolen grocery cart. He was organized. He had systems. He called his daily rounds "rotating on the land." And his rotations covered much of Cabrini-Green's seventy acres.

J. R. would sell to the guys stationed in front of the building or in the lobby. They wanted the new Nas or Wu-Tang CD, a copy of *Space Jam* or the Adam Sandler movie that had opened in theaters that weekend, and J. R. had the bootlegs at cut-rate prices. He'd tease them, saying that he earned more in a week than they did, and he was pushing product that at least was kind of legal. Then he climbed the high-rises, going floor to floor. He knew most people at Cabrini-Green, at least by nickname, and there were regular customers whose apartments he always hit. Other residents, hearing his shouts, beckoned him to their doors. He knew when the checks for welfare, social security, and the city payroll arrived, and he might leave a high-rise with $300 in sales. He'd restock in his Vandura, and then move on to the next building and then the next one. He went through the Whites, the Reds, the rowhouses. He set up shop as well on the floor of JJ Fish & Chicken, a fast-food restaurant on Larrabee, spreading out a mat and covering it with merchandise. The owners accepted the arrangement as part of doing business in the neighborhood.

It added up to a considerable haul. J. R. wasn't exactly the type to report to the IRS, but one year, when he was caught up in a criminal case, his lawyer persuaded him to file a tax return. He listed self-employment earnings of $87,900. He bought new Compaq and Hewlett-Packard computers, investing in software and drivers to make his work copying music and movies easier. He rented his sister Marzetta a place on the South Side, so he could turn her Cabrini unit in 1017 N. Larrabee into a warehouse and workshop. It wasn't just the money that drove him, though of course he chased that. There was the sport of it, the competition, the compulsion to win. There was another hustler, known as Big Boy for his considerable girth, a family friend who'd given J. R. his start duplicating movies. J. R. took over Big Boy's buildings at Cabrini, undercutting his prices. Willie Sr. had to come back from Alabama to intervene, scolding J. R. for his lack of restraint: "Big Boy got to eat, too, son."

And still it wasn't enough. On weekends, J. R. rented tables at the Swap-O-Rama, the giant flea market that operated on a barren rail yard blocks from the old stockyards. He'd scour the market for goods he might resell back at Cabrini-Green, picking up sports jerseys and car sound systems. All those women who did hair out of their apartments—J. R. would buy a case of twenty-four $1 tubes of gel for $6, offering it for $16 to the lady in 1230 N. Burling who specialized in finger waves. For a while he made a killing at Cabrini reselling bed linens and comforters.

Whenever a policeman accused J. R. of piracy, he would rattle off the laws he'd memorized: "These goods are protected under the US Copyright Law, section 117, dash a, dash two, which states that it is not an infringement for the owner of a digital file to make another copy if said new copy is used as a backup." Usually, the cop would move on to the next knucklehead who didn't have a defense lined up. On his rounds, J. R. would even drop into the station house in the 365 W. Oak high-rise, and officers on their shift would pick out toys for their children, horror movies, or porn. Once, when J. R. started selling little plug-in scooters, miniature bikes whose speeds topped out just under the legal limit requiring a license and registration, a police sergeant said he was going to arrest him, for aiding and abetting drug dealers at Cabrini-Green. He accused J. R. of giving criminals the means to zip away from the police. It was exactly the sort of slow-pitch-softball comment that J. R. couldn't help but swing at. "Do you even read a book?" he'd shout at the cop loud enough for the crowds that invariably gathered. He was roused by a sense of indignation as well as by a need to perform. "Do you have any idea what you're talking about?" What about Sammy's Red Hots on Division Street? Sammy's sold hot dogs to dope dealers, right? And did the officer know that Mr. Gyros down the block also provided criminals with the sustenance to carry out their illicit activities? Guys in the hood were notorious for shopping at City Sports. Weren't gangbangers using Nikes to run away from the authorities? "I'm too smart for you," he'd tell the police officer, mouthing the words with exaggerated slowness. "This one on the plantation learned to read."

J. R. and Donna, the mother of his four children, had been on and off for seven years, and they'd been off for months in 1997 when she felt moved to reconcile after rewatching the Whitney Houston and

Angela Bassett movie *Waiting to Exhale*. She looked for J. R. and found him in his apartment with someone else. They argued, Donna picking up a fork from the kitchen counter and ramming it into his chest. They were done as a couple. But after a week, he and Donna turned cordial, trying to coordinate as much as they could to raise their children.

That's when he was introduced to Iesha. Eighteen, tall, and graceful, she had a job in one of the Arab-owned groceries along the mall on Larrabee. J. R. liked that she was work oriented, like himself, and that she'd been an athlete, also like himself. When they first started dating, the Master P movie *I'm Bout It* had just released. In no time J. R. sold out of his first fifty pirated copies; by that evening, a hundred other people had asked him for the tape. "When something goes viral on the land, in a project thick with people, you could be a millionaire off it," J. R. roared. He bought ten more VCRs to speed up production. Iesha not only didn't mind holing up in his apartment as he made the VHS copies but was eager to help. That was a woman for him. Their relationship would end bitterly ten years later, as Iesha finally, tearfully, came out as a lesbian. In the nineties at Cabrini-Green, it was hard to admit you were gay to others, let alone to yourself.

But J. R. and Iesha lived happily enough for the time being. "We were the stars of the projects. A hustler and a worker," he would say of their union. "That was my heart." He bought a new Oldsmobile, midnight blue with gold-flake paint, and they would drive to Wisconsin or Michigan for the weekend. Once for Iesha's birthday he chartered a boat; they were served a candlelit dinner on Lake Michigan as they sailed to Michigan City, Indiana, where Iesha dropped another $800 at the premium outlet stores. "I loved flashing money and helping people out," J. R. would say. He got himself colorful Coogi sweaters and throwback basketball jerseys. He burned through thousands of dollars betting on horses at the nearby off-track parlor, getting fat on the cheese fries.

When Iesha gave birth to their first of four children together, they moved into a fifth-floor unit in one of the red high-rises at Cabrini. And there they enjoyed a version of a traditional family life. J. R.'s children would wake him in the morning, clambering atop him with big smiles. They'd sit down for breakfast, drinking orange juice together. He walked them to school, only a couple of blocks away. And then J. R. would head off to work, rotating on the dwindling land.

ANNIE RICKS

WHEN ANNIE RICKS first moved into 660 W. Division, a woman who lived on the third floor of the high-rise stopped her. "You're going to be my neighbor," she said. "Yes, ma'am," Ricks replied. It turned out they were related. Over the years, when the elevators weren't working and Annie had to walk up the five flights to her apartment, she often popped in at this cousin's place on the third floor and sat for a while at the kitchen table to talk. Then she'd walk up one flight and stop again on the fourth floor, where another set of relatives lived. Annie thought about leaving Cabrini-Green after a shooting in the building. But it was the third-floor cousin who soothed her. "Baby, stay," she said. "It's going to get better."

And it always did. Her son Deonta was named the eighth-grade valedictorian at Schiller, where the students would still see Jesse White from time to time. "It's Mr. White to you, and Jesse to your mama," he'd say when they shouted his name. Annie's other children won basketball titles and all-tournament honors, and their trophies lined her living room. Her older children found jobs working in construction, retail, and a private residential complex built atop the old Madison Street Skid Row, an area that had been renamed the West Loop. In one of the classrooms in which Annie worked, the teacher surprised her with a pair of earrings as a thank-you. Her rent fluctuated, from as much as $300 a month when the adults on the lease were all working, to less than $50 a month when they weren't. One

Christmas, her son Reggie had his name pulled out of a bag, and he was "adopted" by the Slick Boys, the rapping cops bringing him gifts for the holidays.

When he got older, Reggie would say that Cabrini-Green was never as bad as people said. "They maintained that it was the most notorious project. But the two big things that people say about it are Dantrell Davis and Girl X, you feel me? They hold on to that." Reggie had gone to school with Girl X, as she was called to protect her identity. In 1997, the nine-year-old was attacked on the sixth-floor landing of her Cabrini high-rise, just feet from the apartment in which her grandmother had lived for forty years. She was raped and then choked with a T-shirt, the assailant spraying roach repellant in her mouth and leaving her for dead in a pile of dirty snow. She survived, though she was left blinded, mute, and partially paralyzed. Again, another unimaginable crime at Cabrini-Green dominated headlines, and for outsiders, at least, the tenant lawsuit that was holding up development came to seem obscene. "What lifestyle is it that residents want to protect there anyway?" a columnist for the *Tribune* asked.

The city had already invested tens of millions of dollars in public funds into the surrounding blocks for upgraded sewers and utilities, for the widening and repaving of roads. Outside the Rickses' home along Division, a 145,000-square-foot shopping center was under construction. Work had also gotten under way on a new library and new police headquarters. Seward Park was renovated, and a manufacturing district was being revived on neighboring Goose Island, with huge tax incentives for companies willing to relocate there. Across the street from Seward Park, the Oscar Mayer factory had closed for good in 1992, after a century in the neighborhood. But in 1997, Old Town Square opened on the site, with 113 new condos and single-family homes. Dan McLean, the well-connected developer of Central Station, Mayor Daley's new home, secured the desirable land and lucrative tax breaks in exchange for setting aside a dozen of the units for CHA families.

Eager to settle its lawsuit with residents, city officials tried to assure Cabrini families that the rebuilding in the neighborhood was in their best interest. Weeks after the Girl X assault, the mayor's office held a community meeting at one of the local schools. A special assistant to Daley showed the crowd "before" and "after" slides of the plans for the area. "Not a pretty picture. . . . Not much to look at," he said of the "before," which for everyone else was both past and present. The "after," he promised, would be a "complete community." A Cabrini resident cut him off: "A complete community for who?" They already had a community, people shouted. At another meeting, after the same official invited residents to take part in the planning process, a woman told him that her forty-nine years on the Near North Side had taught her one thing. "It's always the same. The forums start, then it breaks down," she said. "They'll go behind your back." A Daley official at one of these public meetings chastised the crowd for interrupting her before she could deliver prepared remarks. A Cabrini tenant shot back: "You interrupted a way of life, lady."

The Coalition to Protect Public Housing was formed in 1996 by Carol Steele, a Cabrini-Green resident, and Wardell Yotaghan, from Rockwell Gardens on the West Side. The group hoped to safeguard residents from these dubious development plans, taking as its motto "Redevelop! Don't Displace." The state of public housing might have been deplorable, but the Coalition argued that the solutions so far didn't look much better. "It wasn't about tearing down the bricks and mortar; it was about moving out the people," Steele said. She hosted Saturday workshops and town halls at Cabrini, telling residents about studies that tracked Section 8 families in Chicago and found they were ending up in all-black neighborhoods of concentrated poverty. Many landlords refused to rent to large families with children altogether, and a third of the people given Section 8s were unable to find suitable housing before the vouchers expired. The Jewish Council on Urban Affairs, the Chicago Coalition for the Homeless, the Community Renewal Society, and other civic-minded nonprofits joined the coalition and its fight to preserve affordable housing in the city. A pamphlet the group distributed asked, *"IF NOT HERE . . ."* above a photo of a boy standing outside a seven-story réd Cabrini high-rise, followed by the question, *"WHERE?"*

The group's largest action, "a people's march to protest public housing policies," took place on June 19, 1997, Juneteenth, the holiday celebrating the day in 1865 when slaves in Texas and other Southern states learned belatedly about their freedom. Some two thousand people gathered outside city hall and the CHA's downtown offices. Many of them had come to Grant Park just two days earlier for the rally celebrating the Bulls fifth title in seven years. "This championship goes to all the working people here in the city of Chicago who go out every single day and bust their butts to make a living," Michael Jordan had announced from the stage. For the public housing call to action, there were speeches, performances, and booths to register voters. Those gathered sang an old spiritual that had been used throughout the civil rights era and which took on special meaning when uttered by people fighting to stay in their housing of last resort. "Like a tree that's standing by the water / We shall not be moved." They needed to stand together and demand a more inclusive city. "If you don't plan your community's future," a Coalition handout warned, "someone else will!!"

Annie Ricks knew about the meetings and marches. She heard her neighbors talk about the city's schemes to shut down Cabrini-Green. She didn't really believe it, though. She pointed to 1230 N. Burling, the high-rise across the field from her. The building had new doors and security cameras and two elevators that worked. After school most days, Annie brought her students over to play at the building's brand-new playground. "Why would you tear down a building when you just put money into it?" she said. After the Girl X incident, Ricks saw that there were more news crews and police officers in the area. But that also felt to her like more of the same. On her way across the parking lot to Schiller school, she was stopped by a cop. He demanded to know where she was going. "Work," she said. "Where are you going?"

One day she told Ernest Bryant that she was ready to get married. Although he had never been on her lease, they had thirteen children and had been together for a quarter century. "Either you marry me, or get out," she said in her half-joking way.

"All right, Jeffery," he agreed. "I'm going to marry you." He called his mother, who said Annie was already her daughter-in-law. They had a big wedding at a banquet hall on the South Side, holding the service and reception at the same place so people would have less trouble getting there. Annie's aunt officiated, and her children got drunk. "We had a good time," Ricks said.

14

Transformations

I N THE SUMMER of 1998, officials from HUD called Mayor Daley's office. The federal government had controlled the Chicago Housing Authority since taking it over in 1995, and in those three years, by several measures, the agency had improved. Joseph Shuldiner, the HUD executive charged with cleaning up the CHA, had previously led the authorities in New York City and Los Angeles. He arrived in Chicago to find Section 8 files packed away neither in chronological nor alphabetical order; no new vouchers were being issued despite 48,000 families languishing on the waiting list, some for more than twenty years. Most of the employees he inherited seemed to be associated in some way with different aldermen or state representatives, and the companies with CHA contracts were almost all politically connected. "Everything was political blood sport in Chicago," Shuldiner said of his introduction to the city. He hired a private management company to oversee the voucher program, and he whittled down the CHA staff, eliminating hundreds of jobs. Rather than the housing authority attempting to manage its own properties, or turning to residents to do it themselves, Shuldiner outsourced these duties to private firms at nearly every development. A new system was implemented to track work orders and service requests. For the first time in a decade, an audit showed that the agency's financial records were complete and presented fairly. With these changes in place, the federal government was ready to give the CHA back to the city. The thing was, Daley wasn't sure he wanted it.

The mayor was keenly aware that his family name—unfairly and not—was tied to the legacy of public housing in Chicago. (Upon sitting down with Shuldiner, Daley read him the minutes from the 1959 US Senate hearing in which the first Mayor Daley said he didn't want "high-risers" for his city.) Daley wasn't interested in incremental upgrades at the CHA. He didn't believe public housing developments could ever be assets to the neighborhoods around them, and he didn't want to go to Washington with his hands out every year begging for another allowance of HOPE VI funding to make improvements around the edges. If the lawsuits at Henry Horner and Cabrini-Green had taught him anything, it was to avoid some piecemeal change, hung up by litigation at every development and then shackled by each ensuing consent decree. The city's public housing projects were monuments to failure in so many different forms—civic and individual, political and historical, physical and economic. Like their place in the city's psyche, their solution needed to be monumental. "Make no little plans," Daniel Burnham had declared about his design for Chicago at the dawn of the twentieth century. Daley said he would take back the CHA only if he could tear down all of its high-rises. "'This is the end of your political career. You do this, this is impossible,'" Daley said he was warned. "'It's all African American. You're not going to solve this. Don't do it.' Sure, everybody believed that. Leave it as is. But you wouldn't have a city of the future, you'd a city of the past."

In September 1999, Daley presented what would be his most sweeping urban renewal program. Under the Plan for Transformation, as the endeavor became known, the city would demolish every remaining public housing family high-rise, knocking down some 18,000 units. Over ten years, and at an estimated cost of $1.6 billion, Chicago would build public-private mixed-income developments on the cleared land and rehab existing low-rise public housing. The total would add up to 15,000 new or renovated family units, plus an additional 10,000 for senior citizens, reducing the current stock of 38,000 down to 25,000.

Daley cast the undertaking as an even more radical transformation of Chicago and its denizens, saying that neighborhoods too long under the pall of towering public housing would finally be imbued with vitality and reconnected to the rest of the city. The very landscape would be remade, the skyline altered, the street grid restored. The replacement housing would be built on a "human" scale, with only a third of the apartments going to CHA residents, thus breaking up the old concentrations of poverty and triggering further commercial and residential investment. And the mostly black residents who had lived in these neighborhoods in social and economic isolation would now reap the rewards of the prospering city. "I want to rebuild their souls," Daley declared.

Julia Stasch helped pull together the mayor's public housing transformation strategy, first as Daley's commissioner of housing and then his chief of staff. Discerning and conscientious, she recognized that there was little research into whether families of different economic classes would live amicably side by side in the replacement buildings, or how to manage properties with both private condo owners and public housing renters, or how exactly the poor would benefit from their close proximity to the gainfully employed, or even what the right mix in these buildings should be, or, more significantly, what would happen to the vast majority of public housing tenants who didn't get into the more sparsely populated, deconcentrated buildings. "Mixed-income buildings are not the answer to low-income housing," Stasch would say rightly. But she thought it anachronistic to look to government-run public housing to make up the brunt of a city's overall portfolio of affordable options. "The cavalry was not coming," she would explain. "The only way to get public benefits funded today is by harnessing the market motives of private entities." This meant that in addition to mixed-income buildings, vouchers would serve an ever-increasing number of families, and former high-rise residents would have to be relocated to rehabbed public housing complexes that were less monolithic in their original designs. There were also single-room occupancy hotels—one would be built next to the new shopping center on Division and Clybourn. And under Clinton, the federal government was producing hundreds of thousands of affordable units through low-income housing tax credits, which offered incentives to private developers to reserve a small percentage of apartments in their buildings for people with modest incomes above the public housing threshold.

Stasch had served as a deputy in the Clinton administration's General Services Administration before coming to city hall. (Asked on her federal employment form if there was anything in her past that could be an embarrassment to the president or the country, she wrote that she'd lived a counterculture life in San Francisco in the 1960s; she wasn't asked to elaborate.) In thinking about the Plan for Transformation, she wrestled with the same concerns that went into the reform of the welfare system and that had always bedeviled public housing. Should the subsidy go only to the neediest, lowest-income families? For how long? And how to refresh it for other needy families, discouraging recipients from staying in the system for generations? She wasn't sure of the answers. What she felt unequivocal about, though, was the harm caused by concentrated poverty. "What's the antidote to that?" she said. "It's mixing people from different economic backgrounds." A pragmatist who believed that the perfect is the enemy of the good, she refused to let public housing families remain in the dangerous status quo of high-rises. "The bad situation in public housing required that we end it."

The financing of the public-private real estate deals under the Plan for Transformation involved a Rubik's cube of moving parts—tax credits, soft loans, city and state funds, developer fees, HOPE VI grants. Various public and private entities had to be satisfied and countless rules abided or officially rewritten. A large chunk of the funding for the Cabrini-Green redevelopment was expected to come

from the sale of market-rate apartments. But the project also benefited from $280 million in tax-increment financing—a tricky bit of
temporal juggling in which city money is made available for improvements in the present based on the presumption of the future gains
in real estate taxes. To pull off the colossal endeavor, Daley, Stasch,
and the rest of his team said they needed to be free of numerous
federal strictures, and they asked for special dispensations on allow-
able construction costs, contracting guidelines, rent limits, admission
requirements, and many other rules. HUD had a designation called
Moving to Work for its highest-performing agencies that allowed for
greater flexibility. Chicago was still anything but high performing, yet
that was the classification the city now sought. Andrew Cuomo, the
future New York governor, was then Clinton's secretary of housing,
and thought submitting to Chicago's demands would set a terrible
precedent. If he made these exceptions for Daley, then the mayors
of Indianapolis or Cincinnati would be in the next month asking for
similar deals. But Daley went over Cuomo's head to President Clinton.
The CHA was named a Moving to Work agency, granting it dozens of
waivers of federal provisions.

"We are going to be fighting this tooth and nail," Carol Steele
announced. The Coalition to Protect Public Housing called for a
moratorium on any demolition until the city could demonstrate that
the Plan for Transformation wasn't just scattering families into the
unknown. Steele had taken part in the years of negotiations that had
finally settled the lawsuit Cabrini-Green residents filed against the
CHA back in 1996. Under a consent decree, signed in 2000, six of
the red high-rises would be demolished, not eight, and the replacement housing would include three times as many public housing units
as Daley's proposal had allotted. Tenants would also be able to form a
nonprofit subsidiary and take 50 percent ownership of a partnership
with the developer of the site; they would share in fees and profits, the
money going to aid both displaced and remaining Cabrini residents.

Steele now pointed out what she saw as basic arithmetic errors

in the Plan for Transformation. The city was guaranteeing families
with a valid lease as of October 1, 1999, the right to return to rehabbed public housing, or a new mixed-income development. But
Daley's plan would reduce the city's stock by a third. In recent years,
the city had eliminated thousands of public housing units, and many
more sat vacant. Yet there were 56,000 families on the CHA waiting
list, 24,000 waiting for a housing choice voucher, and 80,000 people
who were homeless. And many more low-income renters in the city
struggled to find affordable housing options. Where would these tens
of thousands of families go? "The mayor says that no one is going to
be displaced," a protestor at a rally outside city hall said. "We say the
mayor is a liar."

In a way, the CHA was hoping to replicate the success of a scatter-
site housing initiative started as part of the *Gautreaux* case, the class-
action housing desegregation lawsuit against the agency that was first
filed in 1966. Rather than waiting on the CHA to build housing in
racially diverse or "revitalizing" areas, a judge finally ruled on the
case, in 1976: families from Chicago public housing would enter a
lottery, and the winners would be moved with vouchers into private
rentals in better-off neighborhoods. By the late nineties, 7,100 families had taken part in the program and been relocated to areas with
less crime, improved schools, and better job prospects. Only a few
hundred were moved a year, so receiving neighborhoods—often white
and suburban—didn't feel inundated and the relocated families could
receive counseling and assistance with mobility, and be checked on
and studied over time.

But now the city was hoping to find permanent and temporary
homes in the private market for more than twenty thousand additional low-income families. A study of the Chicago rental market
that was demanded by the Coalition to Protect Public Housing
showed that the city wasn't even close to being able to handle this
flood in demand. Chicago had an overall residential vacancy rate
of just 4.5 percent, among the tightest in the nation, with the avail-

ability largely concentrated in the poorer sections of the city. More than 40,000 families were already renting with Section 8 vouchers, but another half million renters in Chicago qualified for the subsidy, and the city had an affordable housing deficit of 140,000 units. The numbers gave the federal officials funding the project pause. "HUD cannot approve a plan in which the CHA will displace more families than the housing market can accommodate," the department wrote in a letter to the CHA.

In January 2000, Daley went ahead with the big public rollout. He'd convinced the Central Advisory Council, the tenant leadership body representing all the city's public housing developments, to endorse the proposal, breaking with the tenants who were trying to stop it. "With or without us, they were going to submit it anyway," Francine Washington, the president of the group, said. "At least this way, we are able to include our comments." With the tenant group backing the plan, HUD signed off on it as well. Then the city's largest foundation, the John D. and Catherine T. MacArthur Foundation, threw its institutional support behind the mayor's redevelopment effort, funding dozens of grants and funneling some $65 million into the efforts.

Stasch had been hired by MacArthur to lead this charge. She set about recruiting the many partners across the city needed to pull off the endeavor. She funded the researchers to study its efficacy, and she enlisted businesses to employ CHA residents who would now need to meet more stringent work requirements. "We were able to have an impact on a huge swath of the city that had been negatively impacted, and on the lives of thousands of people," Stasch said. "It was a once-in-a-generation opportunity."

J. R. FLEMING

ONE NIGHT IN May, three years into the Plan for Transformation, J. R. Fleming was leaving the Cabrini rowhouses where he'd just sold an electric scooter when he saw the flashing blue lights of a police car behind him. He was in the passenger seat, with Iesha driving one of the Vanduras and the children strapped into the back. J. R. didn't have the van registered or insured, but that didn't stop him from jumping out and yelling, "What'd I do?" He wasn't surprised to be pulled over. A couple of months earlier, police officers had raided his sister's apartment in 1017 N. Larrabee, where J. R. stashed his bootlegging equipment and his merchandise. Acting on a tip, officers found several computers, seven VCRs, televisions, speakers, and video cameras—equipment, they claimed, that was being used to duplicate the stacks of unlicensed music, movies, and video games that they also confiscated from the apartment. J. R.'s lawyer argued that he was a legitimate music and video producer, and there was no proof he intended to sell any of the merchandise. When prosecutors asked for a warrant to search the hard drives of J. R.'s computers, they were denied. The charges were dropped, though J. R. never saw any of his confiscated equipment again. A week later, he found dents kicked into the doors of one of his vans and the side-view mirrors bent back. He was told the cops did it.

What surprised J. R. about being pulled over that night was how the officer responded. He was a white cop, a mountain, six feet four, 240 pounds. Without a word, the officer reached back and smashed a fist the size of a tree burl into J. R.'s face. Then he hit J. R. a second and a third time, knocking him to the ground. Iesha charged out of the van, yelling that the cop should just arrest J. R., he wasn't resisting. Knowing he was going to be locked up, J. R. tried to hand the cash he had in his pocket to Iesha. That's when the cop crossed the line. He stiff-armed Iesha in the chest, pushing her away.

This was Cabrini-Green on a warm spring night. A hundred people were hanging out in front of the buildings. They saw the cop shove Iesha, and at once there arose a collective "Oooh!" *Was that a punch? Did he touch her titties?* J. R.'s own thoughts cleared. He took a steadying breath. He knew so much in these situations depended on not panicking. Turning to face the officer's partner, who'd followed him out of the squad car, J. R. raised his hands. "He didn't even have to do her like this," he said. Positioning himself with his back to the big cop, he drew a bead on the man's chin. "This could have been avoided." On the last word, he slammed back his right elbow, connecting with the officer's jaw. A tooth arced through the air and landed silently on the ground. The cop followed, dropping like a felled oak. And with that, J. R. knelt down and crossed his hands above his head, trying to give the second officer no more excuse than he already had to shoot him dead. "That was murder she wrote," J. R. said. "That changed the rest of my life."

The arrest didn't come easily. Responding officers arrived, running toward J. R. armed with nightsticks. What was he to do? He defended himself. In the scuffle, he was hit with a steel flashlight. Officers blasted pepper spray into his face. He was wrestled into the back of a squad car and charged with two counts of aggravated battery of a police officer, one count of resisting arrest, and a single count of criminal damage to a government vehicle. As the arrest report stated, "Offender using his feet kicked out passenger side rear window." J. R. said he needed to break the cruiser's window—he couldn't breathe after all the pepper spraying. If found guilty of all four counts, J. R. could spend the next sixteen years in prison. He'd be forty-six when he got out, his children grown.

J. R. had a fairly extensive rap sheet, with eleven misdemeanors for criminal trespassing, theft, disorderly conduct, failure to disperse, and the one previous felony for the drug sale he did up on the far North Side when he was eighteen. But his lawyer argued self-defense, showing that the police had a track record of intimidating J. R. Brother Jim, writing the judge, attested to J. R.'s "complicated relationship with police officers. Some like him and even purchase merchandise from him. Others have arrested him for a number of minor offenses." More than forty Cabrini residents admitted to seeing the incident, but only three of them who weren't family to J. R. agreed to testify, the others fearing retaliation from the police. Officers gave conflicting testimony about who approved the use of excessive force, and a commander's signature appeared on a permission form even though he wasn't on the scene. J. R.'s lawyer requested a history of complaints and internal investigations into misconduct for each officer. The trial dragged on, with continuances and court dates stretching into months and then more than a year.

J. R. believed it would help his cause if he could demonstrate his dedication to his community. He needed to show the judge that he wasn't some thug. The women running his old building allowed him to turn an empty apartment in 1017 N. Larrabee into a recording studio. He brought guys together from the different sections of Cabrini-Green, some of them in rival gangs, to cut a track for the Coalition to Protect Public Housing. "It's such a shame, we living in vain / They took our neighborhood, they don't want us to change / This where my mother grew up / This where my family grew up / Thank the Lord he sent us to save the Greens through us." He took part in 100 Men Standing, a mentoring program for Cabrini-Green fathers. He started a chapter of the Hip Hop Congress, a social activism and arts nonprofit found mostly on college campuses; his was the only one in a public housing development, and he said it was a way to use music to reach the masses, to connect colleges with the community. One of his first events was a Hip-Hop Holidays for the Homeless. "Homelessness ends with hope," he announced. With less time for peddling, he used his heat-press to design T-shirts for the "Save Cabrini" movement.

J. R. STARTED organizing as well for the Coalition to Protect Public Housing. He sat down in Carol Steele's office, where posters displayed words of inspiration: POVERTY IS THE WORST FORM OF VIOLENCE —Gandhi; THE TEST OF OUR PROGRESS IS NOT WHETHER WE ADD MORE TO THE ABUNDANCE OF THOSE WHO HAVE MUCH; IT IS WHETHER WE PROVIDE ENOUGH FOR THOSE WHO HAVE TOO LITTLE —Franklin Delano Roosevelt; TRUST GOD. In a lilting voice that belied her toughness—a bullishness not unlike J. R.'s—Steele explained all that had been going on with the Plan for Transformation. Few city agencies could have handled a task of such immensity, of such political and human nuance. But the CHA had a long track record of being among the least efficient and worst managed of government departments. For patronage hires, it was a backwater, the better skilled of the well connected usually opting to earn a paycheck elsewhere. When the Plan for Transformation got under way, moreover, the CHA was just coming out of its federal takeover, its workforce being shrunk from 2,500 employees to a mere 500. Even for the remaining employees who were diligent, thoughtful, and caring, they were unprepared for the political pressures of the transformation process. The word from city hall was to empty the high-rises, and observers reported a wartime atmosphere at the CHA. "We just really give up on the ones who have a lot of problems," said a contractor hired by the CHA to assist residents with their relocations. "By that time, we just can't do nothing for them. I mean, we can't even find them! So how can we serve them?"

The CHA was surprised to learn how many people in its buildings had mental or physical disabilities, suffered from trauma, or abused alcohol or drugs. These families needed the help of a social worker, not a relocation counselor. "Once we started digging, we had no idea how bad it was from the perspective of people needing services," said Lewis Jordan, one of several heads of the CHA who would be in charge of the transformation. Of the tens of thousands still living in public housing, many had spent decades there, and without the proper guidance, a great number of them didn't see announcements, missed meetings downtown, and, on hearing about the multiple steps involved in their moves, gave up. Those who did follow each task met with relocation managers and attended workshops on good neighboring and money management. They filled out the numerous forms stating whether they preferred a permanent or temporary move, whether they were willing to rent in the private market with a voucher or wished to stay in a rehabbed public housing development. If they were behind on rent or had unpaid utility bills, or if they had a relative on their lease with a criminal record or didn't meet new employment requirements, they raced to become "lease compliant." Then they joined trips to scout homes in other neighborhoods. But the relocation counselors each handled more than a hundred cases at a time. They were supposed to show every family five possible residences. But that didn't happen. The counselors were paid per case closed and in many instances had established relationships with specific landlords.

Overseeing a vast network of thousands of Section 8 landlords presented its own challenges. Many landlords in the private market offered homes far superior to public housing. But there were also many who were inexperienced as far as tying their low-income renters to social services. Some illegally turned families away or refused to repair leaks or faulty furnaces. Apartments weren't up to code. Negligent landlords were rarely punished. And when they were, tenants suffered, since they were forced to relocate again, uprooting their families, starting the search process all over and dealing once more with utilities, the phone company, and a rental deposit.

By the end of 2002, the city had demolished 6,900 units of public housing. But it had yet to rehab any public housing units that weren't reserved for seniors, and it had built only 130 units of public housing in mixed-income buildings since 2000. Many residents, failing to find apartments before their move-out dates, doubled up with family members elsewhere or left their condemned public housing and moved into other public housing in the city not yet scheduled for demolition. Residents too often left one unsafe or unsanitary apartment for another and then another. Not infrequently families were sent across gang boundaries. "The Plan for Transformation was probably the most disgusting thing I ever saw as an organizer. How little care was taken for the people who were being displaced," said Jim Field, who worked first with the Community Renewal Society and then the Chicago Coalition for the Homeless. "The general public actually cares about homeless people. That is not the case with how they feel about public housing residents. The stories coming out of public housing of shootings and killings, most mainline people say, 'Well, we're paying for that.' But they can frequently identify with homelessness."

Facing criticism, the CHA agreed to let an independent monitor assess its progress. Thomas Sullivan, a former US attorney, confirmed that the buildings were closed in an unnecessary rush, that tenants lacked proper counseling and were thrown into a state of mass confusion. Nearly every relocated family, he found, ended up in neighborhoods that were primarily African American, and three-quarters in areas that were overwhelmingly poor. These were not the soul-rebuilding opportunities that the transformation promised, and now families were in strange territory and without their former support networks. "The result has been that the vertical ghettoes from which the families are being moved are being replaced with horizontal ghettoes, located in well defined, highly segregated neighborhoods on the west and south sides of Chicago," Sullivan wrote. The Plan for Transformation didn't so much break up concentrations of poverty as move them elsewhere, making them less visible to the rest of the city. A class-action lawsuit was filed on behalf of the relocated residents, asserting that the city had been resegregated.

The CHA followed many of Sullivan's fifty-four recommendations. It hired a new social service provider, vastly improving how the agency relocated families. But problems with relocations also persisted. Tenants had signed a "right to return" agreement stipulating that anyone with a lease at the start of October 1999 had a shot at coming back to whatever replaced the old high-rise housing projects. Now nine in ten families said they wanted to return to their longtime neighborhoods, twice as many as the CHA had predicted. At Cabrini-Green, the CHA signed a right-of-return contract with 1,770 families. Those who wanted to come back had to take temporary vouchers and move elsewhere. Some didn't meet the new rules for returnees regarding work requirements. Others were evicted with "one strike" drug laws or for other reasons. As more time went by, some died, others were simply lost. The process was much worse at other large developments that were torn down hastily. A few years into the transformation effort, the CHA placed ads in local newspapers, stating that it actually couldn't account for some of the 16,800 families it had ostensibly been tracking, and would anyone from the following list of 3,200 families please contact them in the next ninety days or risk losing the right to return to public housing.

"It was perfect timing," J. R. would say. He was fighting his own legal case as the Plan for Transformation was rolling out. Getting involved was an easy choice. "I decided I wanted to play a role. I saw the suffering wouldn't stop."

AT THE SAME time that the city was carrying out its Plan for Transformation, the housing authority considered selling 1230 N. Burling to its tenants. The building had long been celebrated as the exception to everything believed to be wrong with Cabrini-Green. During the federal takeover of the CHA, HUD determined that the tenant managers of 1230 N. Burling had "improved living conditions" and reduced "neglect and abuse," allowing for "resident empowerment and economic uplift." Now the residents proposed buying their home. Their offer was far below the market value of the property, and they asked also that the federal government continue to subsidize maintenance costs at the high-rise for a decade after the sale. Residents pointed out that developers were being given sweeter incentives to build around Cabrini-Green and that the government paid Section 8 landlords to house low-income renters. By operating 1230 N. Burling as a non-profit co-op, they would be able to provide homes to 134 public housing families. They'd earn "sweat equity" in the property as well, their work compensating for what they couldn't afford in a down payment. The sale was endorsed by local politicians and civic leaders and approved, albeit unenthusiastically, by the CHA. Many deemed the proposal a poor use of public resources with a high likelihood of failure. The Illinois Housing Development Authority and

Chicago's Department of Housing and Economic Development ultimately killed the deal.

Still, the CHA continued to see the Burling building as an ally in its plans for Cabrini-Green. In 2002, Carol Steele ran to unseat Cora Moore as the head of the tenant council for all of Cabrini. When the polls closed, Moore led 260 to 66. But four hours later, an additional 222 ballots from the rowhouses were discovered—and then subsequently lost—with Steele winning 201 of them. After other ballots were disqualified, Steele was declared the winner, with an overall edge of 261 to 214. The CHA didn't want to concede the election to Steele. In addition to the questionable ballots, Steele had been involved in the lawsuit that halted work on the Cabrini Extension high-rises and was a leader in the Coalition to Protect Public Housing. The agency called for a new round of voting. There'd been other irregularities as well: poll watchers were unable to enter buildings where votes were cast; electioneering occurred at polling sites; in one high-rise, voters each received two ballots; and in 1230 N. Burling, Moore's stronghold, the ballots weren't properly initialed. The election ended up in court, with the CHA providing Moore an attorney. In their arguments, the lawyers cited both Shakespeare and the recent precedent of *Bush v. Gore*, which differentiated between the counting of spoiled and missing ballots. The CHA lawyers attacked Steele's character, saying she hired her boyfriend to manage the rowhouses. The back-and-forth lasted eight months, until a judge decided that the transgression of the missing ballots was less egregious than the failure of initialing them, at least as far as Illinois election law went. Steele was declared the winner.

Other buildings at Cabrini-Green also formed their own self-management corporations. The CHA accused the resident management group that Steele led in the rowhouses of permitting squatters to stay in empty units and of spending some $300,000 on salaries and fees, in excess of what federal law allowed. Then in 2003, at 6:00 a.m., the CHA raided 1230 N. Burling. Before padlocking the door of the resident management office, CHA officials emptied everything out, tossing computers and files into Dumpsters. Dolores Wilson's daughter Cheryl found hundreds of dollars' worth of tenants' uncashed rent checks scattered by the wind. In a press conference, the new head of the CHA, Terry Peterson, said he was firing the resident managers. He displayed photographs that had been taken during an earlier inspection of the building. They showed garbage in piles and a shopping cart lying on its side in a hallway. "Our residents deserve to live in safe, clean, and well-managed buildings," Peterson said. "Those are the kinds of conditions that are unacceptable and that we're going to hold our property managers responsible for cleaning up."

The tenants of 1230 N. Burling, assured for years of a special status, struggled with their comedown. "They didn't give us a reason, they didn't tell us who they were," Cannon said. "They just said, 'Your services are no longer needed.'" He'd been in charge of maintenance of the building for more than a decade. Like the other resident managers, he lost his job. The building wasn't perfect—nearly a decade after its renovation, the front doors were dented and the Plexiglas in the entranceway fogged. On the landings and ramps, years of winter ice and salt had stripped the concrete floors, and the brick walls in the hallways, painted a stygian red, were scribbled with graffiti. But it wasn't what the CHA was claiming. The conditions *before* the residents took over were truly unacceptable. "Our suspicion is obviously that they want to take over the building, deplete the number of people in it, and then say it's uninhabitable and then move people out so they can demolish it," the head of the Chicago Coalition for the Homeless, Ed Shurna, said.

"I felt it was the end," Dolores Wilson said. "If they could just come in and close down a corporation, they could do anything." She expected to hear from the politicians and officials who had worked with her and her neighbors over the years, who had trained them and funded their efforts. "I was thinking Bertha Gilkey would have come to cheer us up or tell us how to follow through. I thought she would have at least called," Dolores said. "She knew we were doing a good job. No one ever gave an explanation. You would think they would tell us why they did that to us."

J. R. FLEMING

J. R. DISCOVERED that activism was a lot like peddling. He made the same rotations on the land at Cabrini-Green, visiting all the remaining high-rises and rowhouses. His customers still listened to his playful patter when he was selling tales of injustice and broken promises. He told residents that the city's "Plan for Devastation" could be deemed a success only if the metric were getting poor people off prime real estate and moving them to areas where there were even fewer jobs and transportation options, where crime, gangs, and schools were all worse. He liked the sport in being a gadfly, a pest to those in power. With the other members of the Coalition to Protect Public Housing, he'd go to CHA board meetings, using the public comment portion to call out officials as "chump change opportunists" and "sell-out Daley lackeys." His voice booming, he repeated how Cabrini residents had been displaced, how their rights of return were being hindered, how resident-owned businesses didn't receive contracts and tenants weren't hired for construction jobs.

The CHA launched a new brand reinvention, executed for free by the Chicago ad agency Leo Burnett, turning the letters of the housing authority into the word CHAnge. The image "This Is CHAnge" appeared all over the city alongside pictures of residents and their testimonials of better lives. In 2005, artist-activists covered up the ads on bus stops with similar looking displays that reworked the agency's logo to "This Is CHAos." The alternative campaign featured large photos of either Mayor Daley, Terry Peterson, or other housing officials alongside accusatory questions: "Are tourists more important than the poor?" "Do money and politics mix?" The one featuring the prominent Chicago real estate developer Dan McLean read, "When

the Mayor's Plan for Transformation called for the demolition of
Cabrini-Green, McLean got the scoop early and began buying up
adjacent properties. . . . McLean cashes in on our tax dollars with
no risk." J. R. defined the CHA's new slogan as "Chicago Helps All
Negroes Go Elsewhere." He advised tenants who were facing removal
or bouncing around among Section 8 rentals. It was exciting to lead
a hundred residents in chants as they marched downtown. "Aren't
y'all tired of running?" he demanded of his neighbors. "We're from
the Greens. We can't let them take our history."

Back in 1998, for the fiftieth anniversary of the United Nations'
Universal Declaration of Human Rights, Carol Steele and other CHA
residents had traveled to New York to give testimony about their fight
back home. Cabrini-Green was so vilified, so closely associated with
crime and drugs, that it was hard for residents to find allies in their
hometown. But in New York, activists from around the world made it
clear that they considered the battle on the Near North Side to be part
of their larger global struggle. For J. R., the formulation of housing
as a human right provided him a game plan that he could run with.
He started carrying a copy of the Universal Declaration of Human
Rights in his back pocket, whipping it out and brandishing it like a
weapon. In the same way he used to spout copyright laws when hawking bootleg gear, he now recited the UN document, article twenty-five
dash one: "Everyone has the right to a standard of living adequate for
the health and well-being of himself and of his family, including food,
clothing, *housing*, and medical care and necessary social services."

He began to attend human rights conferences and events across
the country, with organizations paying his way. Every couple of
weeks, J. R. went before the judge overseeing his assault case and
asked her for permission to leave Illinois—to attend conventions in
Atlanta, Philadelphia, New York, Virginia, New Jersey, Maryland. In
a letter Steele sent to the judge, she praised J. R.'s work, writing, "He
encourages young men and women to participate in the processes that
will determine the fate of their community as the Chicago Housing

Authority's Plan for Transformation changes the nature and makeup of the Cabrini-Green community." The judge was both impressed and confused. She looked at J. R.'s past arrests, the crimes he'd committed, his lack of even a GED, and she tried to square that with the letters inviting him to speak at the United Nations World Urban Forum. She permitted all the trips, even one to Caracas, Venezuela, where J. R., sporting a Che Guevara bandana under a White Sox cap, met with Hugo Chavez's government about supplying oil to poor Chicagoans. The judge also let J. R. go to Las Vegas, where he and Iesha got married, a last-ditch effort to save their relationship.

J. R. led a busload of Cabrini residents down to New Orleans as well to assist with relief after Hurricane Katrina in 2005. He returned repeatedly over the next three years, observing each time how the city was restored everywhere but in its poor, black neighborhoods. In New Orleans's Lower Ninth Ward, the wrecked homes with spray-painted Xs sat unchanged; cars remained where they'd been washed under porches or flipped on their sides. "It was so obvious in New Orleans," J. R. would say. "Once the water went down, you could see exactly where the rebuilding occurred and where it didn't. The demographics of the whole city changed. That 'tools of displacement' shit is real." Shortly after the flood, the New Orleans City Council voted to demolish much of the city's remaining public housing. Six of its largest complexes were shuttered. The city lost more than three thousand previously occupied units of public housing. "We finally cleaned up public housing in New Orleans," a US congressman from Louisiana boasted. "We couldn't do it, but God did." For J. R., New Orleans clarified what was going on in Chicago. Chicago didn't need a flood. He and his neighbors were being pushed out of the city center, out of Daley's global city. He was going to push back.

Despite the objections of the dozens of police officers who attended the hearings, the judge in J. R.'s case ruled that he would be sentenced to no jail time. She gave him community service and two years' probation, and ordered him to attend anger-management classes. "I've got anger issues with the system," J. R. would tell the psychologist. He grew his hair out and twisted it into ringleted dreadlocks. It was around then that he legally changed his name from Willie McIntosh Jr. The new name was part of his new identity. "I'm Willie J. R. Fleming, human rights enforcer," he'd say by way of introduction. "The 'J' is for Just, the 'R' is for Righteousness." James Martin, the plainclothes cop at Cabrini-Green nicknamed Eddie Murphy, had known J. R. and his family for decades. Reflecting on J. R.'s personal transformation, he joked that his colleagues on the police force had messed up. They should have left the young man alone when he was just peddling DVDs and tube socks: "Now they went and woke him up."

15

Old Town, New Town

KELVIN CANNON

When he was young, Kelvin Cannon believed the towers of Cabrini-Green were as immutable as mountains. They were as much a part of the natural landscape as the boundless plains of Lake Michigan. "But I got older," he said, "and I got wiser." He watched as the Ogden Avenue Bridge, his childhood playground, disappeared, the barricaded overpass eventually torn down. He learned for himself the history of Cabrini-Green—that the neighborhood had previously been Irish, Swedish, and Italian; that the Cabrini rowhouses and high-rises were built on top of an existing ghetto that had once seemed permanent until it wasn't. Cannon came to subscribe to a pragmatism born of fatalism: change was something the powerful imposed on everyone else. In 2003, Mayor Daley had ended debate over the future of a downtown airport by sending in bulldozers late at night and tearing up the runways; he had Lake Shore Drive moved to create a downtown museum campus. Imagine what he'd do to public housing. "Cabrini was an eyesore for the Gold Coast," Cannon said. "The plan to transform us was mandated in the early seventies. It was only a matter of time, inevitable." The community might be able to save the rowhouses, he believed. The high-rises, no chance. "I was taught it's best to sit at the table with people, so we can coexist," he explained. "All we can do now is negotiate and try to get as much as we can coming back as public housing units."

Cannon respected Carol Steele. He didn't deny that the lawsuits and all the protests had served a purpose. But Cannon figured that Ms. Steele had lost her way. A decade had passed since the millions in federal HOPE VI dollars were first earmarked for the rebuilding of Cabrini, and in all that time little had been done. For the last three of those years, Steele was president of the Cabrini tenant council; she'd tangled with the city and with Peter Holsten, the developer who won the contract to rebuild parts of Cabrini-Green. "Three years wasted," Cannon said. "In the process of tearing down, we were supposed to be rebuilding the mixed income. But Ms. Steele couldn't come to an agreement with Holsten. Nothing was built. A lot of residents were being displaced, and a lot of those people were never found again."

In 2004, Cannon decided he'd try to unseat Steele as tenant council president. He considered himself more than qualified. At forty-one, he'd lived at Cabrini-Green for all but the three and a half years that he'd been incarcerated, and it had been nearly a quarter century since his sole conviction for armed robbery and fifteen since he'd left the Gangster Disciples. He'd apprenticed under Cora Moore when she was tenant president. He'd worked alongside her and Dolores Wilson to self-manage their high-rise. He'd done maintenance and construction. Steele's vice president reported Cannon to a gang crime commission, saying that a Gangster Disciple was trying to take over the multimillion-dollar rehab of Cabrini-Green. But Cannon let it be known that he'd turned himself over to God. Given a second chance, he'd dedicated himself to the community.

In some conservative circles, in fact, Cannon was hailed as an exemplar of the redemptive capabilities of self-reliance. He was invited to join members of the Empowerment Network, a profaith policy group that included Pennsylvania senator Rick Santorum and other socially conservative members of Congress. As a former gang member, Cannon was presented as living proof that even the intractable problems of the inner city could be overcome through the "resilient power of freedom and free markets, and the reliance of its citizens on our deep wellspring of faith in God." The Empowerment Network's director, David Caprara, had specialized in resident-directed initiatives at HUD under Jack Kemp. "God bless your achievements, your family, and 1230 N. Burling," he wrote to Cannon. "You are our hero!"

Cannon wasn't about to run away from his gang past, either. He wagered that people at Cabrini-Green would recall that he'd commanded the Disciples without abusing his authority. Being a GD governor, he felt, proved that he was someone who stuck with whatever he started and rose to positions of leadership. He put on a shirt and tie and a V-necked sweater, donned his leather coat, and started to walk the land at Cabrini-Green, going door-to-door to every apartment, telling people why he should be president. He collected signatures and asked residents for their vote. Cannon liked to canvas alone. He felt people would recognize his fearlessness and determination. He made the rounds of the Whites, the Reds, and the rowhouses. When he completed the circuit, he started again. Half of the units were now vacant. But nobody really knew which ones were empty or occupied, who was on lease and could vote and who wasn't. For three months he tried to reach everyone. Cannon spoke in intermittent bursts, his words suddenly rushing out in a torrent as he tried to explain that the election was a referendum on redevelopment. Under Carol Steele, they were missing their only chance to be a part of the evolving area. They had to embrace the change. They had to trust that too many lawyers were already involved in the Plan for Transformation for it to go wrong.

Carol Steele and her backers called Cannon a crony of the CHA and the developers. They pointed out that when Holsten opened North Town Village, a private mixed-income development across from 1230 N. Burling, he'd given Cannon a job there as a security guard. Steele said Cannon was at best naive. Hadn't a decade of broken promises proven that the CHA couldn't be trusted? How many times had residents been told they were part of the planning process only to be pushed aside when the city decided to go ahead with its own agenda?

Steele wasn't holding up the process, she said time and again; she was trying to make sure that what occurred was legal and in compliance with its stated aims: "We said, 'Show us the land. Show us the money and show us the housing.' But the CHA couldn't produce. That's why it's been a long, drawn-out process." She wanted proof that those with a "right of return" could truly return to Cabrini-Green, that they would actually be mixed into the new housing. Earlier in 2004, Cabrini residents filed another lawsuit against the CHA, after four hundred families were issued eviction notices that would have pushed them out long before the building of any replacement housing. Experience had taught her that the CHA fulfilled its promises to tenants only when a judge ordered it to do so.

On the day of the vote, Cannon passed out walkie-talkies to his friends and family to coordinate as they ushered residents to the polling places and watched for improprieties. Turnout, as at most public housing elections, was light. Also, many who did vote spoiled their ballots—they liked both candidates, so they picked each of them. In the final count, Cannon received a couple dozen more votes. Steele challenged the results, but the CHA approved the tally. She said she lost because the CHA and the city wanted her out; Cannon wasn't so much elected as selected. Cannon credited his victory to a higher power than the Chicago Housing Authority and even Mayor Daley. He said he had God on his side.

ONE OF CANNON'S first acts as president was to move the tenant council offices north of Division, into 1230 N. Burling. He combined two units on the first floor of his high-rise to create a space big enough. Then, weeks later, he signed the agreement to finally start work on the new housing that would replace the Cabrini Extension towers. He put his faith in Peter Holsten, who'd been awarded the contract to develop the mixed-income complex. "We are blessed to have him as a partner," Cannon said.

Holsten had grown up in the western suburbs, and after an MBA at the University of Chicago took him to the South Side, he stumbled into real estate and won business with the city. Public housing, he related, "came to encompass my life work." He said he learned quickly that it wasn't enough to hand public housing families the keys to an apartment and say, "Live happily ever after." He believed in a model of in-your-face management, strict but respectful: "'I'm sorry, Mrs. Jones, you're being put out because your son set fire to the apartment. Maybe we can get your son in a program somewhere, and I can convince management to keep you here.' That's the yin and the yang," Holsten would explain. "Obey the rules or else." He hired case managers to help tenants search for jobs and stay current on their bills; he kept a social-service consultant on-site in each of his properties. "I want all the kids in my buildings to go to college, and I want the heads of household to be employed and for the cycle of poverty to be broken," Holsten said. "I know that's not going to happen. What do I settle for short of that? I try to give them the best house that I can. I try to get everyone to be neighborly."

Over the previous decade, a dozen market-rate developments with public housing units mixed in had opened on the periphery of Cabrini-Green. Private developers competed for the building rights as well as the tax credits and favorable financing. For centrally located Cabrini-Green, it didn't take much to lure investors. Residential property sales in the two-block radius around Cabrini-Green totaled less than $6 million in 1995, but five years later, at the start of the Plan for Transformation, annual sales had reached $120 million, and from 2000 to 2005, total sales neared $1 billion.

In 1999, when Holsten won the rights to develop North Town Village, he priced the for-sale units to sell, at a discount of 15 percent below the area's going rate. Seventy-nine of the 261 condos and townhomes were reserved for public housing families, and he thought buyers might balk at the unique arrangement. Part of the pitch made to young professionals was that they would be "social pioneers," embarking on a new form of urban living. People of different races and classes were going to live alongside one another, under the same roof. But being a pioneer also meant acting boldly and getting in ahead of others; it meant staking a claim before prices rose and the land was fully settled.

Months ahead of construction on North Town Village, when a sales trailer first opened at the site, the units went fast: forty-seven that day, eighty by the end of the week. Cabrini towers still defined the landscape. But buyers trusted that the high-rises would soon give way to even higher property values, and North Town Village was pronounced one of the city's hottest real estate markets. "Just a stone's throw from the Gold Coast, River North, Old Town, and Lincoln Park, Cabrini-Green is easily one of the most sought-after neighborhoods in all of Chicago," gushed a write-up from a local real estate firm. North Town Village townhomes sold for as much as $475,000. A young couple who bought a condo described a buying frenzy, one woman yelling at her husband, "Just buy whatever's left!"

The area that had been Cabrini-Green—and before that Little Sicily and Little Hell—some Realtors now tried to rebrand as "New Town," dropping the Cabrini-Green moniker inimical to their interests. Peter Holsten named the mixed-income complex that replaced several red Cabrini high-rises along Division Street "Parkside of Old Town." Among the dignitaries at the groundbreaking for the complex, Holsten was joined by the CHA's Terry Peterson, Alderman Walter Burnett, and Kelvin Cannon, his shovel raised, wearing a do-rag beneath his hard hat and a white cashmere topcoat despite the mud. Daley announced that Cabrini-Green was now "part of the larger Old Town neighborhood."

Latasha Ricks, one of Annie Ricks's daughters, helped build Parkside's combination of mid-rises and townhomes. "Even though I was president and a partner in the development," Cannon said, "I helped build it, too. I was out there in the fields. I was there as a laborer." During the construction, Holsten presold 70 percent of Parkside's market-rate units, purchasers putting down 5 percent of the price. That was in 2006, two years before anyone could move in, and buyers paid half a million to nearly three-quarters of a million for the townhomes, with the condos starting at $300,000. Many people snapped up two units, figuring they'd flip one later as prices continued to climb. Holsten planned initially to complete one mid-rise before starting on the others. But with the money from presales and the increasing demand, he began construction on the entire 760-unit complex, all 228 townhomes and several mid-rises.

One of the new owners, in 2006, was Abu Ansari, whose partner, Mark, surprised him with the floor plans to their unbuilt Parkside condo. A stage actor from Texas, Ansari had moved to Chicago in the early nineties, and he'd followed news of the Plan for Transformation, wondering whether public housing residents would be forced to leave at the very moment that their neighborhoods started to flourish. His own mother had grown up in public housing in San Antonio. And as a black gay man in a relationship with an older white man, he was attuned to the stark divisions of race and class in his new city. How many times had he used the word "gentrifier" as a slur? And now he might be one of them, in of all places Cabrini-Green.

But two years is a long time to get over your misgivings. Prices around Cabrini rose higher still, and Abu and Mark talked about their prescience for getting in early. And when their Parkside mid-rise actually went up, the design exceeded their expectations. The building was the opposite of the towers it replaced. Squat and broad, it was a multisided prism, reigning genially over the corner of Division Street and Seward Park, like a cross-legged Buddha. Its orange-bricked facade was adorned with splashes of purple ornamentation and decorative pillars. Every unit had a balcony. Abu and Mark started to embrace the idea of themselves as pioneers in an exciting social experiment.

When Abu and Mark finally moved in, it was bliss. Their ninth-floor apartment was roomy and modern, facing out over the new supermarket complex and Lincoln Park beyond. They hosted dinner

parties, proud to show off their new home. At soirees set up by the
building's management, the owners met one another, the conversation
turning invariably to their foresight and collective good fortune. The
public housing units had yet to be filled, but the condos were nearly
sold out. The owners could watch from their balconies as potential
buyers for another mid-rise under construction next door lined up at
a sales trailer.

This was in the early fall of 2008. A month after Abu and Mark
moved into Parkside of Old Town, the global investment bank Lehman
Brothers filed for bankruptcy. For more than a decade, as home prices
soared, people had embraced a faith preached by government and
business alike—a global prosperity gospel—that the housing market
could only go up. The Plan for Transformation was devised amid this
real estate bubble, and it was premised on the sale of market-rate
units at the new mixed-income complexes. Without these sales, the
public-private developments couldn't fund construction costs or guarantee loans for the large capital projects. Public housing was created
in the 1930s because the for-profit real estate market, by its very
nature, was unable to provide decent and affordable homes for Americans at the lower rungs of the economic ladder. And seventy years
later, the speculative market had again made its limitations plain.
People at Cabrini-Green weren't thinking about the deregulation of
the banking industry, or credit default swaps, or negative amortization
loans, or collaterized debt obligations, but the building of replacement units came to a halt at every Chicago public housing development. Construction on the second phase of Parkside of Old Town
ceased. With the trucks and the men in fluorescent yellow vests gone,
the poured foundation sat exposed like a Roman ruin, harking back
to an age that had yet to be.

The Parkside buyers who hadn't finalized their purchases simply
walked away, giving up their 5 percent deposits. Peter Holsten lost
half of his presales. His market-rate partner on the project, a national
developer with properties across the western states, filed for bank-

ruptcy. Holsten listed Parkside apartments at half their peak price, with only a 3 percent down payment required. With the help of the MacArthur Foundation, the city created an incentive program called "Find Your Place in Chicago," paying $10,000—and for a limited time only, $15,000—to anyone who purchased a unit in a mixed-income development. It offered grants to buyers to cover closing costs. When Holsten couldn't pay back a $32 million Parkside construction loan to JPMorgan Chase, he and the bank discussed taking the unsold units to auction, allowing Chase to write off the remainder of the loan as a loss.

For the Plan for Transformation and Mayor Daley, that was unacceptable. Cabrini-Green was too notorious to fail. The city bailed Holsten out. He was due to receive a public subsidy of several million dollars only after a certain number of Parkside units were sold. Although the collapse of the housing market left him nowhere near that threshold, the city gave him the money, allowing Chase to renegotiate his loan. A couple of months later, the city financed about half of the $42 million cost to build another eight-story Parkside building on the corner of Oak and Larrabee that would include thirty-nine units of public housing.

Most of the owners in Abu and Mark's building were underwater, owing more on their condos than they were worth. Those who bought multiple units, hoping to flip them, were stuck with two apartments they couldn't sell. Neighbors lost their homes to foreclosure. Parkside buyers had been celebrated as urban pioneers and risk takers. Now they felt like suckers. Owning was supposed to mean you had the right to exercise choices. Buy, sell, move, take out a home equity loan. You wanted to make money, to see a return on your investment. That's what they imagined as the American way. Never mind that the federal government devoted three times as much each year to mortgage interest deductions and other homeowner subsidies—essentially public housing for homeowners—than to the entire annual budget of the Department of Housing and Urban Development. But now the housing market had failed them as well. The remaining Cabrini-Green high-rises across the street no longer were so easy to ignore; they loomed ominously, like giant tombstones.

That was when the public housing families began to move in. Housing officials had talked passionately about the prospect of "productive neighboring" in the mixed-income developments, the families in public housing networking with and learning from their middle- and upper-class neighbors, working adults modeling a professional lifestyle for lower-income youth. Even in the best of times, it seemed starry-eyed to imagine that interactions would happen in these buildings across race and class and age that rarely occurred elsewhere in the city. But amid the worst financial crisis in eighty years, the divisions became starker. Parkside included on the same floors not only people of different races and classes but also the conflicting American ideologies of self-sufficiency and social obligation, of home ownership and public assistance. Stuck with their bum mortgages, though, the owners came to resent the people who were "living for free" beside them in nearly identical apartments. Obsessing over the number of units that sold and at what reduced price, they complained that public housing families were "taking over."

At their meetings, the condo owners talked endlessly about "situations" in the building that were depressing their property values. They'd seen renters come downstairs to get mail in their pajamas. That wouldn't look attractive to a potential buyer. They proposed a rule prohibiting pajamas in public spaces. They fretted over the congregating of public housing families in the lobbies. They had to be frank: a large gathering of young black people in the entranceway looked more like the old Cabrini-Green than a new one. Someone suggested removing furniture from the lobbies. Someone else seconded a rule restricting the size of gatherings inside apartments, to keep noise levels down. They complained about public housing residents who left the gates to their townhomes open—it was ugly and a hazard for anyone walking a dog. One woman suggested a ban on garden gnomes. She was afraid no one would buy a place after seeing a gnome.

Ansari was especially sensitive to the rising tensions. There were other black condo owners, but the fault lines formed mostly around race. He tried to befriend the older woman who lived next door to them. A former Cabrini-Green resident, Ms. Smith told him how much she loved her Parkside apartment, and spoke proudly of her son who was attending college. She also played her television at an unbearably high volume, and music with a heavy bass thudded while Abu and Mark tried to sleep. When one of them knocked to mention the noise, Ms. Smith apologized and promised to keep it down. But in no time the music again thumped and the television again blared, and the requests and the promises were repeated, with increasingly less politeness on both sides. To make matters worse, both Abu and Mark were laid off in the downturn. Ansari was able to find part-time work. But Mark remained at home. Every day, at the same hours, he soon discovered, Ms. Smith's adult children visited, often with their own children in tow, and within minutes the family members were screaming at each another. Mark could set his clock by it. He'd try to ignore the yelling, the music, the TV, but he couldn't even think. When he went over to complain, Ms. Smith sometimes bemoaned the sorry state of her life, telling Mark that she had nothing and he had everything. Mark couldn't believe it. He was fifty and out of work, with a new mortgage to pay. By then, he'd been unemployed for more than a year, and he worried if he'd ever be hired again.

On a night when Mark phoned several times to complain about the racket, one of Ms. Smith's sons threatened him, calling him a "faggot." Mark had the son banned from the building. The son who'd been away at college returned. "Please, please," he begged of Ansari. "I'm trying to keep the family together." Eventually, though, the manager of the building stepped in, moving Ms. Smith to a different Holsten development. Mark was overjoyed to see them gone. Ansari's feelings were more mixed. "I felt relief but also a deep sadness," he'd say. "My biggest nightmare came true." As a black homeowner, he'd forced out a returning Cabrini-Green resident. The Smiths certainly weren't uplifted by their brief time in Parkside of Old Town.

The economy began gradually to improve. Mark found a full-time job. Slowly, the empty Parkside condos sold. You couldn't beat the location, at the crossroads of major boulevards and bus and train routes, right by the Loop and the burgeoning tech sector forming nearby in River North. The hulking Montgomery Ward warehouse hugging the river just a couple of blocks away had been remade with the use of city subsidies and now included high-end condos, corporate headquarters, restaurants, a spa, and a yacht club with boat slips onto the water. Ground was broken on another Parkside of Old Town mid-rise, as well as on several other private residential buildings in the immediate vicinity. Demolition of the remaining Cabrini-Green high-rises proceeded. The CHA renter who moved into Ms. Smith's apartment was a widowed grandfather who kept to himself. Every once in a while, Abu and Mark could hear as if at a distance the tinkling of contemporary jazz. They worried mostly that they were too loud for him.

DOLORES WILSON

MOST OF THE high-rises at the city's other large public housing developments were already leveled or in the process of being cleared. It seemed to defy reason that the most infamous of Chicago's gallery-style public housing complexes, the one on the primest real estate, would outlast them all. But the lawsuits at Cabrini-Green had stalled demolition. Now it raced to catch up. Five white high-rises were razed over thirteen months starting in December 2005: 1340 N. Larrabee, 630 W. Evergreen, 714 W. Division, 534 W. Division, and 624 W. Division. Kelvin Cannon had lived in two of the buildings as a child. Then came 1121 N. Larrabee and 1159–1161 N. Larrabee, two of the Reds, and in a flurry, in 2008, five more of the red high-rises, a total of 538 apartments. Dolores Wilson watched as one by one the white high-rises around her were crowned with the rooftop billboard that heralded their demise—HENEGHAN WRECKING WE MAKE SPACE. And then she couldn't look away as the space was made. The red-cabbed cranes came in, a wrecking ball the size of an ocean liner's anchor swinging into the top floors, the prefabricated facade crumbling like old chalk. Another crane with steel teeth spat water as it tore into a wall, exposing someone's room behind it. Bit by bit the buildings disappeared, one apartment, a bank of them, several floors. The sheared towers revealed dozens of brightly painted rooms, like a box of pastels.

When the letter arrived saying 1230 N. Burling would be next, the high-rise facing it was in the process of being reduced to a couple of twenty-five-foot mounds of rubble. A group of residents in Dolores's building met in the first-floor rec room to decide what to do. One of the elected tenant reps warned everyone about the CHA's scare tactics. Kenneth Hammond, a forty-one-year-old lifelong Cabrini resident, had been living in an adjacent high-rise, and when it was torn down he'd leaped as if from a sinking ship to the Burling building. Hammond had watched as his former neighbors took whatever far-flung addresses the housing agency assigned them. They were made to believe they didn't have a choice, and most of them ended up in neighborhoods that seemed at least as poor and segregated and violent as the homes they'd left. "Don't think it's near safe out there as it is in your own community," Hammond cautioned. Cabrini-Green had its problems, of course, but it was better to face those problems in your own home than on someone else's turf. Living at Cabrini, most people had figured out which stores offered credit, what church, social service agencies, community leaders, and neighbors could be counted on in a bind. If someone had trouble with rent, a towed car, or a family member who'd been arrested, she at least knew where to find help. In the private market, you were on your own. "Who can you go to?" Hammond said. "Once you in these new communities, they going to shut you out. Only thing they going to tell you, 'Man, Shorty, go back to where you used to stay.' The CHA trying to set us up to fail."

The last thing Dolores wanted to do was move. At eighty-one, after a half century at Cabrini-Green, she had her church, her friends, her family, all on the Near North Side. Her youngest son, Kenny, had passed away a couple of years earlier, of pneumonia. But both of her daughters lived in 1230 N. Burling as well, Debbie on the same floor and Cheryl two floors up, on the eighth. Dolores appreciated that people wanted to fight to stay. "Carol Steele works her butt off," she said. "Praise the Lord for her. I'm glad she doesn't give up." But in her conscientious way, Dolores attended the relocation meetings; she filled out the surveys and paperwork and followed up with a case manager.

She had zero interest in testing the private market after all her years in public housing. She'd heard about families renting with vouchers who'd been forced to relocate two and three times. She didn't think she could handle one move, let alone more. She walked across Division Street to Parkside of Old Town and submitted an application for a unit. When she visited one of the apartments there and stepped out onto the balcony, she lurched back. The fence between her and a six-flight fall was no more than waist high. "I got conditioned where I can walk up to a fence and lean against it, even if it looks like a prison," she explained of the ramps in the high-rises. But she knew Peter Holsten and respected the way he ran a building. A woman at the Parkside office told Dolores, "We'll let you know." But no one ever did. She turned in an application at the Cabrini rowhouses as well. At least she could have a front and back door, a little yard. But the manager there blew her off, telling her to come back Friday and then didn't show himself. Eventually, a representative from the CHA told her to choose somewhere else. She joined a scheduled tour of the rehabbed public housing across the city, a carload of residents leaving from in front of her building.

Dolores liked what she saw at Lawndale Gardens, a rowhome development on the West Side, in the Little Village neighborhood. But then she found out it was a block away from the Cook County Jail, with its ten thousand inmates. "I can't be close to a jail," she shouted. It would have been like someone who believed in ghosts living along-side a cemetery: she'd have heard the suffering of the inmates at all times. They drove her next to the Dearborn Homes, on the South Side, once a part of the four-mile corridor of public housing that stretched south from the Loop along State Street. Dolores was tired of running around, so she took it. Her unit there would be a tiny one-bedroom, on the fifth floor of a nine-story building. The CHA set it up so Cheryl could relocate to an apartment on the same floor.

Even for someone as meticulous as Dolores Wilson, the actual move was confusing. She had so much in her four-bedroom apart-ment to pack, and so little space in her new place to put it. She was told she would be given boxes for all her belongings. Then on the day of the move, the men from Big "O" Movers said there weren't any more boxes and they had to move her right away. Dolores cried as a lifetime of mementos went into the trash. She lost every letter she ever received. She lost her wedding photos and pictures from her trips to Jamaica with Hubert and documents from her time as head of the resident management group. Hubert had received more than twenty certificates and awards, for coaching sports teams and running his drum and bugle corps. The Corsairs had won twenty-six trophies. Dolores had honors from HUD, the Chicago police, the wa-ter department. All of that history was gone, every plaque, tossed in a city dump somewhere. "I hate Daley Jr. with a passion," Dolores would say. She blamed him for doing nothing when he was Cook County state's attorney and the news broke about Jon Burge, the Chicago police commander who supervised the torture of more than a hundred black men. She blamed Daley for caring more about the land at Cabrini-Green than the people who lived there. And she blamed him now for the loss of her belongings.

KELVIN CANNON

When KELVIN CANNON'S first three-year term as tenant council president ended, the CHA extended it to five years without holding an election. With the foreclosure crisis and the Plan for Transformation stalled, the agency must have felt justified in ignoring this isolated little fief in the democratic process. Cannon didn't question why two years were added to his term. He just went ahead and did his job the best that he could. He met with developers to go over the constantly reduced scope of their plans. Hundreds of landlords in the city renting to Section 8 families were also losing their buildings to foreclosure, sending their tenants back into the private housing market at a perilous time. Many of these Cabrini families were now calling their old home, asking Cannon for assistance. J. R. Fleming, in his role with the Coalition to Protect Public Housing, sometimes harangued Cannon, accusing him of cozying up to the developers. But all in all, Cannon believed he'd been among the fairest, most loyal tenant council presidents ever at Cabrini-Green. He was proud of the work he'd done. Yet in 2010, when Carol Steele pressed for an election, he decided not to run. With the construction of the first phase of Parkside completed and the rest slowly beginning to resume, Cannon felt it was time to move on.

He'd lived in 1230 N. Burling since 1983, when he was paroled there. In 2010, the tenants were told the building would be torn down, and Cannon moved into an apartment in Parkside of Old Town. It was a two-bedroom, with two bathrooms and a balcony that looked onto his emptying Burling high-rise. The apartment had a sunny open design, hardwood floors covering the connected kitchen and living room, and carpeting in the bedrooms. There were stainless steel appliances and a granite countertop, and it was roomy enough for Cannon to put a weight bench alongside his couch. He hung his panther paintings. His photograph collection covered the walls, the tabletops, and an upright glass display case, the pictures telling his life story—Cannon in a group shot with his children, or alongside his father during a prison visit, or dressed to the nines at the Players Ball, or standing beside Mayor Daley, or with Reginald and William Blackmon when they were teenagers. Cannon's oldest brother stayed in the second bedroom. Cannon's mother and another brother also moved into a Parkside unit. People griped that Cannon had gotten the apartments as payback for his support of the developers and the city. But he dismissed that talk as plain jealousy. He applied for an apartment and qualified just like anyone else. "I can't help that we passed drug tests and background checks, and others didn't," Cannon said. And of course he looked out for his mother. She was eighty. "If I don't help my mother, who will?"

For Cannon, living in Parkside was a blessing. "You're supposed to pass that blessing on to the next person," he said. So he understood why others complained about it. By year ten of the Plan for Transformation and with almost every high-rise gone, only 372 Cabrini-Green families had moved into a mixed-income development. Those applying were excluded from one of the limited spots if they or a family member couldn't pass a criminal background check, if they had unpaid rent or utility bills, if they or any other adult on the lease failed a drug test, or if they weren't working thirty hours a week or their children weren't enrolled in school. Holsten estimated that he accepted one of every five public housing applicants. But it turned out that a small number actually applied. They assumed mixed income wasn't for them. Of the sixty Cabrini families Holsten had asked to compete for twelve units at North Town Village, only two saw the process through. He began working with church leaders and local officials to bring out more potential tenants. He hired William Gates, the former prep school basketball star featured in the documentary *Hoop Dreams* and a longtime rowhouse resident. An ordained minister, Gates conducted outreach for Holsten in the Cabrini community.

Cannon knew some Cabrini families at Parkside who felt that the old public housing, for all of its faults, had been more hospitable.

In their new mixed-income buildings, they had to undergo regular housekeeping checks, and there were fines for such things as riding a bike on the sidewalk or playing loud music from a car. There was a sense you could get in trouble just for being yourself. When anyone on a public housing lease did mess up, and he or she was arrested, even for a misdemeanor, even if the arrest didn't lead to a conviction, the entire family could be evicted. Many times the CHA struck a deal with a leaseholder, saying she could stay only if the arrested family member was barred from ever setting foot in the apartment. You might have felt like you won the lottery when you got into Parkside, but now you had to choose between your daughter who was caught smoking weed and a roof over your head.

In Cannon's building, you'd smell weed coming from some of the condo owners' apartments. They weren't drug tested. They kept dogs and barbecued, but public housing families weren't allowed to do either. Cannon didn't want a dog himself, and he didn't mind the drug tests or the inspections. He believed rules were necessary to keep order, and his market-rate neighbors had paid for certain privileges. Different laws for different tenants, though, didn't do much for fostering "productive neighboring." The public housing families in the new buildings weren't allowed to form tenant councils. But the owners had their condo associations, and they set the rules that affected everyone.

In any condo building in the city, owners were going to be wary of large numbers of renters; renters didn't have savings tied up in the property and had no financial stake in the building's future. That wariness was even greater at Parkside, since the renters paid only a small subsidized rent. Imagining himself an ambassador, Cannon took it upon himself to bridge that divide, greeting what he called his "European" neighbors. He'd say, "Good morning" and "Good evening." Sometimes people ignored him, but if he saw them later in the Seward Park field house or the stores across Division Street, they smiled with a flash of recognition. That was a start. Occasionally people in the building responded by welcoming him to the neighborhood, not considering that they were new and he'd been there for close to fifty years.

ANNIE RICKS

IN NOVEMBER 2010, the woman from the Chicago Housing Authority phoned again, pleading, begging Annie Ricks to *please* move somewhere. She told Ricks it wasn't safe to live alone like that in the projects. Fifteen stories, 134 units, the entire 1230 N. Burling building a pillar of darkness save for seven remaining families. At night the glow from the Rickses' windows in apartment 1108 looked like the distant beacon of a lighthouse. With a laugh, Ricks told the official to let her be. She'd lived at Cabrini-Green for twenty-one years, and she was going to enjoy her Thanksgiving at home.

"You have to find a *new* home, Ms. Ricks."

"No, I don't. I can stay in 1230 N. Burling forever."

One of the tenant reps in the building told Ricks, "I'm not going until you go." But then he left for the rowhouses. "He ran like a chicken with its head cut off," Annie said. When the moving trucks came for the last seven families, only Ricks and a man on a lower floor refused to leave; then he relented, taking an apartment assigned to him.

By that wintry fall, Cabrini-Green's twenty-two other towers had each been shuttered. Ricks was the last tenant of all the Cabrini-Green high-rises. She was also the last remaining resident of any gallery-style public housing high-rise in Chicago. Every tower had been closed, tens of thousands of families packed up and moved elsewhere. She'd outlasted them all. At fifty-four, Ricks was a grandmother nearly forty times over. She'd lost all her teeth except for two on the bottom, and her black hair was streaked with white. "I'm the last woman standing!" she liked to proclaim.

A CHA relocation specialist had taken Ricks to see Lawndale Gardens, not far from where she grew up on the West Side. But she

had memories of coming home from Harrison High School and running from the Mexican and Puerto Rican students who threatened to
beat her up. She didn't want that for her children or grandchildren.
She went to check out the Cabrini rowhouses as well. "Doesn't this
area seem right to you, Ms. Ricks?" a housing official asked her. It
didn't. Of the 600 homes there, just 150 of them, a sliver along Cambridge Avenue, had been rehabbed. The other 450 had been cleared
of residents in 2008 and left to sit, even though the Plan for Transformation promised that the entire area would be fixed up and repop-
ulated. Column after column of the barracks-style homes remained
empty, their doors and first-floor windows covered in boards, their
postage-stamp gardens gone to field, the perimeters blocked off with
chain-link fencing, like some military outpost long after the soldiers
left for home. Annie knew, too, that the boys in the rowhouses were
warring with the young guys from the "orange doors," the Evergreen
Terrace apartments on Sedgwick, north of Division Street. Two of her
former students had been shot on Cambridge.

A pro bono lawyer suggested that Ricks agree to move into the
best of the replacement homes being offered. But she decided she
wouldn't accept anything less than a four-bedroom apartment, which
was what she was eligible for. She wasn't illegal in her apartment.
She paid her rent on time. No one caught her with drugs or guns.
So how could the CHA just assign her a new apartment? There
were six people on her lease: three of her children—Reggie, Rose,
and Raqkown—two of her grandkids, and herself. For twenty years
Ricks had lived in a four-bedroom in a neighboring high-rise, 660
W. Division. Then in 2008, when that building was about to be torn
down, movers with the housing authority carried out all her belongings and set her up in 1230 N. Burling. In her Burling apartment,
she had five bedrooms and a big living room, with a closet that fit an
air mattress for whenever another family member needed to spend
the night. She'd bought a brand-new washing machine from Home
Depot less than a year before, thinking she'd be able to use it in her

apartment for years to come. If she had to move, she wanted a unit that also had a washing machine hookup. "Why lie and say they don't have a four-bedroom at the CHA?" she'd complain, citing a litany of dates and detailed conversations she'd had with housing officials. "I know they have a four-bedroom. They have so many four-bedrooms over there." She cut herself short, a look of surprise giving way to a wide, closemouthed smile. "But I can't never be mad."

Alone in 1230 N. Burling, she decided to turn the solitude into a celebration. In her apartment she cranked the music all the way up: "Power & Praise" on AM 1390. No one else was left in the building to complain. Her daughter Rose and several of her grandchildren jumped rope in the apartment. The younger ones hula-hooped and pogo-sticked. Ricks fired up the grill and barbecued on the open-air gallery. An icy gale sliced through the fencing, eleven stories high, but she repeated to anyone who would listen that she didn't mind one bit.

As much as Ricks professed her love of living alone in 1230 N. Burling, it took its toll. One Sunday she returned from church to find a river gushing outside her apartment. She grabbed a broom and attacked the water, whisking it off the ramp and elevator. She traced the source into the vacant apartment next to hers, the stream leading into one of the bathrooms. A bathtub had been stopped up, the taps turned on full blast. "Sabotage!" she cursed. Now they were trying to flood her out.

And it wasn't easy clambering up the stairwells, hiking those eleven stories. She wasn't young like she used to be, and the elevators didn't always work. The elevator was often broken in 660 W. Division. But there she lived on the fifth floor and could drop in and see family members on the way up. At 1230 N. Burling, nobody else was left to invite her in for a visit.

On December 1, 2010, the CHA took Ricks to court, seeking an emergency injunction to close her building. The agency argued that it was not only absurd but also an undue financial burden to keep

the heat on in a fifteen-story high-rise with just a single unit occupied. A federal judge agreed. Ricks had to leave within ten days.
Annie said she would move if she could go across the street, into
Parkside of Old Town. But the housing authority deemed it unfair
for the Ricks family to skip ahead of the many Cabrini residents on
the Parkside waiting list. And maybe the agency considered Annie Ricks too obstinate, too intractable, for the delicate balance of
mixed-income living.

They gave her a unit seven miles south, in Wentworth Gardens,
a low-rise public housing development of 422 units sandwiched between the fourteen lanes of the Dan Ryan Expressway and the fenced
parking lots for the White Sox baseball stadium. When she'd toured
Wentworth earlier that fall, she saw teenagers and young men selling
drugs out in the open. She didn't know these young men. At Cabrini,
she had known all the boys who lingered outside her high-rise—she
likely helped raise them in the after-school programs she ran or in the
classrooms in which she worked as a teacher's aide. She'd probably
shared meals with them or their families. Her lawyer explained to the
judge that Ricks was afraid to live at Wentworth Gardens. But the
judge said drugs were being sold all over Chicago. That was that. "I
had to go," she'd concede. "Either that or be homeless."

On the morning of her move out of Cabrini-Green, December
9, 2010, Ricks rose early to take her bath. It was twenty-one years
almost to the day since she'd walked there in a snowstorm. As she
readied herself, Ricks pulled her hair back into a sprig of a ponytail.
She put on a white collared shirt with gray stripes and a caramel-
colored leather coat. Annie and her children had been up most of
the night finishing the packing. On the way out of the building for
the last time, Ricks stuck her head into the management office. Kenneth Hammond and a couple of other tenant representatives were in
there. "Hey, y'all. I'm leaving," she said. She passed by the building's janitors gathered on the ground floor. Her new apartment in

Wentworth Gardens didn't have a laundry hookup. She'd given away her new washing machine to a janitor who'd looked out for her. He and his brother had already carted it off. "I'm-a miss you," Ricks told him. "I'll miss you, too," he said.

As it had the day she arrived at Cabrini-Green, snow had fallen. But this time the reporters really did show up to chronicle her move. They were there from ABC, NBC, CBS, Fox, WGN, the *Tribune*, the *Sun-Times*, the *New York Times*, the *Wall Street Journal*. "An inglorious end to an infamous era," as one of the newspapers put it. Another described Cabrini-Green as "the housing development that came to symbolize the squandered hope of them all." The reporters jostled around the last high-rise tenants of what had been the city's—and really the country's, and the world's—most iconic public housing complex. Rose Ricks, then seventeen, rolled a suitcase with a Route 66 sticker on it. She said, "I've been here basically my whole life. Like it's hard leaving when you've got so much memories of it. You knew everyone. You felt safe."

Men from Big "O" Movers carried out most of the belongings, but Annie didn't trust them with her most prized possessions. Her son Deonta lugged a cardboard box filled with the trophies he and his siblings had won for basketball tournaments and perfect attendance and the one that he took home for being valedictorian of the school that was still there just an empty field away. Schiller had been given a new name and a makeover, and it reopened the year before, serving only those students who tested in, ranking it among the best elementaries in the entire state. Children no longer walked a few hundred feet to get there but arrived by car and school bus. Deonta stopped outside the high-rise. He wanted people to understand that Cabrini-Green was more complicated than they thought. There was the myth of the place as something terrifying. You said you were from Cabrini-Green and people recoiled as if you had a deadly disease they might catch. But living at Cabrini, the Ricks family had experienced the fullness of it. They also had fun there. "There was more good than bad," he tried to explain.

When the trophies were stowed and the furniture loaded onto a truck, Annie Ricks ducked into a sedan the same off-white color as her high-rise. The car spun its wheels on the snow, gained traction, and she was gone.

They Came from the Projects

THAT WAS DECEMBER 2010. By March, the last Cabrini tower had been stripped of every window, door, and cabinet. The steel curtains covering the ramps were yanked off and carted away. The fifteen-story high-rise looked exposed, like a giant dresser without any drawers. On a frigid evening, a hundred people huddled in the field below. They were gathered to commemorate the end—the demolition of 1230 N. Burling would begin in the morning. Behind the crowd, to the east, in the real estate running to the Gold Coast and Lake Michigan, the sunset played off the glass-and-steel towers; above them, the twilight passed through the skeleton of the Burling building.

They could also see a man inside the tower, his movements revealed by the yellow vest he wore and the horn of light extending from his headlamp. There he was, alone in what might have been a friend's apartment. And now he was on one of the upper floors, beneath ceiling fans sent whirling by the wind. His name was Jan Tichy, and he was a conceptual artist and professor at the School of the Art Institute of Chicago. On his commute to the Art Institute in recent years, he'd passed the "infamous" housing project, witnessing the late stages of its erasure. Tichy's art tended to wrestle with architecture as a site of social and political conflict, and he proposed to the CHA a live art installation in the last Cabrini high-rise. It would be a way to memorialize the disappearing icon. To his surprise, officials at the agency loved the idea. For his Project Cabrini Green, Tichy partnered with local after-school programs, and their students contributed poems about the housing development. Jada Jones wrote in "Trapped,"

> *What they called hell,*
> *We had named home.*
> *On the streets where children were once found,*
> *Transformed by evening to brutal battlegrounds*
> *That's where we belonged.*

Tichy recorded the students reading their words, and he translated the syncopations of their voices into flashes of light, blinkering LED circuits that he placed inside 134 ammo boxes painted a bright orange—one box for each 1230 N. Burling apartment. Tichy planned for the lights to run on a loop throughout the thirty-day demolition, or until they were destroyed by the wrecking ball. All of this would be filmed from a camera set up on a balcony of one of the Parkside of Old Town apartments across Division Street. Dual images, along with audio of the children reading their poems, would play simultaneously on a website and at the Museum of Contemporary Art, just a mile east of the building.

It seemed only fitting that the passing of high-rise Cabrini-Green would be turned into art and idea. The development had long existed in both its solid state of concrete and steel and in the vaster realm of abstraction, evoked as shorthand to convey a certain idea of the inner city. Just three weeks earlier, a prime-time network crime drama called *The Chicago Code* aired an episode it named "Cabrini-Green." Ronin Gibbons, a character on the Fox series played by Delroy Lindo, is a Cabrini-Green native who has lifted himself up to become both alderman and syndicated crime boss. (The real-life alderman of the neighborhood, the Cabrini native with a bank robbery in his past, Walter Burnett, joked about any likeness: "You know that character grew up in Cabrini. It's insulting. At least give me some kind of royalty.")

In voice-over, Gibbons describes his childhood at the Near North Side housing project, how people "deteriorated" along with the high-rises, personal squalor matching that of the physical environment. "I made it my mission to see that prison torn down," he says, as the episode chronicles, with surprising accuracy, the many ways the city's public housing went wrong—the lack of funding, the loss of rent-paying residents, the eighteen thousand people living on seventy acres, more than two-thirds of them children. The demolition of Cabrini-Green, Gibbons goes on, "showed me there was hope. Hope that a kid from the streets could rise up and be an instrument of change." As we hear his soliloquy, we watch a murder that Gibbons has ordered on another black kid from the streets, someone who happened to cut into his profits. In the city's churn, the war on anything considered blight, there are always going to be winners and losers. "Today is a great day," the alderman declares. "The first day of a better Chicago."

Some of those at the vigil outside 1230 N. Burling that night shared the sentiment of the fictional alderman, believing that the event marked the end of a failed fifty-year public enterprise, a long-overdue exorcism of inner-city horrors. "This is really the final high-rise," a CHA official who'd joined the ceremony announced triumphantly. "The legacy with the gallery-style public housing, it's a final farewell." Photographers for the National Public Housing Museum—which was still struggling to raise funds to open its museum of "stories" and "ideals" on the site of the former Jane Addams Homes a couple of miles south—were out there capturing the moment. So, too, were crew members from the Cabrini-Green documentary *70 Acres in Chicago*. One of them lowered a boom mic in front of another CHA employee who was describing the deal the agency had struck that week with the retailer Target to build a 190,000-square-foot store on the site of the cleared William Green Homes. He said the Target would bring jobs to the Cabrini-Green residents who remained in the rowhouses and the surrounding mixed-income buildings. But many residents didn't see the trade-off as a net gain. "People got dispersed for a Target?" a tenant leader asked, his voice going high pitched with incredulity. "All of this"—he swept his arms, taking in the expanse of empty land where thousands of people had lived—"for a Target?"

As night fell, Jan Tichy was still moving through the tower, re-checking that the boxes housing the LEDs were fastened tightly to whatever secure things he could find in the stripped apartments—electrical outlets, light sockets, heating pipes. Most everything in the units had been removed, though traces of the lives spent there re-mained. A painting of a black dreadlocked Jesus. An elaborate mural of a giant Bugs Bunny alongside a wizardly Mickey Mouse and a whirling Tasmanian Devil, each of the characters holding blunts the size of their arms, their heads in clouds of smoke and eyes crenellated in red. In a bedroom, a single piece of white paper was still taped to the cinder block. It was a list scrawled in red marker, "The 7 Keys to Success"; among them, "Set and Achieve Goals," "Make Wise Choices," and "BE PERSISTENT." In another apartment a green and gold mural covering an entire wall stated, "I NEED MONEY."

While the people in the parking lot below waited for the start of Project Cabrini Green, the marching band from the Marion Nzinga Stamps Youth Center arrived. The flag-waving majorettes stepped crisply heel to toe, the *tack-tack-clap* of the drums and cymbals ring-ing the air. The dancers spun as a boy on drums shouted a command. Everyone in the brigade dropped low, their bent legs swaying like butterfly wings. Then the LEDs started dancing from the cavities of the high-rise. It was like stealing a peek inside 134 still-occupied homes, nighthawks in each planted in front of a television, the flick-ering sets washing the darkened rooms in their glow.

ANNIE RICKS

FOR A COUPLE of weeks, the media followed Annie Ricks and her family to Wentworth Gardens, eager to tell the next chapter in the life of the last Cabrini-Green high-rise tenant. Annie's new apartment was on the second floor of a three-story walk-up, part of attached buildings that formed a horseshoe around a shared courtyard. During the news interviews, the boys who hung outside tried to angle their way into the camera shots as Rose Ricks shooed them away. "My family from Cabrini-Green made history," Annie would proclaim. "We made nationwide history." But the attention waned, and the reporters stopped calling. Annie's twenty-year-old son, Reggie, liked to tease her, knotting his elastic face as he addressed his mother by her middle name: "You used to be popular, *Jeffery.*"

"I am popular, Reggie *Lee* Ricks," she chided playfully. "God loves me."

Like many of the people who relocated elsewhere, Ricks regularly traveled back to Cabrini-Green. She was one of those gathered outside 1230 N. Burling on the eve of its demolition. Her daughter Latasha drove her, and her son Kenton snapped photos as Ricks watched roosting geese and a rabbit scurry over the cleared fields. On other days, she spent time with family members in the remaining rowhouses. She'd drop into Carol Steele's office or track down Joe Peery in his Parkside of Old Town building. At the end of the school day, her son Raqkown usually took the bus not to Wentworth but to Cabrini-Green, playing basketball at Seward Park. Some relocated parents enrolled their children in Jenner or Manierre, another local elementary, having them ride a bus and train rather than put them in schools in their new neighborhoods. Others met up with friends at the rowhouses, hanging on Cambridge. Better there than an unfamiliar strip of Austin, Englewood, or South Shore. At a town hall meeting for those displaced by the Plan for Transformation, former Cabrini residents talked about the dangers in their unfamiliar neighborhoods. "My grandchildren can't even go outside and play in their so-called new community since they received their Section 8 apartment," a woman told the crowd at Saint Joseph's. "They are having gang wars all around this city when a new face comes on the block, so that is why my grandchildren come back to Cabrini to play with the people they know and the people who know them."

Former residents also got together for "Old School" picnics and reunions. Several Cabrini-Green Facebook pages formed, people reporting job opportunities and business ventures, sharing words of inspiration and announcements of deaths in the Cabrini family. Oftentimes a post showed a photo of one of the high-rises—"Who can say what building this is?"—leading to lists of competing memories. There was nostalgia, but more than that as well. No one forgot that they'd seen a dead body or that the elevator was too often out of order. But they remembered the good, too, more powerfully now that they felt adrift elsewhere.

The Ricks family's first days at Wentworth Gardens were promising enough. With all the coverage about the end of Cabrini-Green and its last tenant, many people recognized them from the news. An older woman, upon meeting Annie, said, "Hey, you're the lady on Channel 7 with that one tooth." "Yes, I am," Ricks responded proudly, and they became friends. Ten other Cabrini families had relocated there as well, including Donna, the mother of J. R. Fleming's first four children. They all welcomed the Rickses. When winter ebbed, Annie barbecued out in the courtyard, a communal feast. At nights during the summer when the White Sox hit home runs, she sat in the window with her grandchildren and pointed to the fireworks from the exploding scoreboard that lit up the sky.

Wentworth Gardens was built in the 1940s on four square blocks next to the White Sox stadium, and like other public housing developments with "garden" in their names, it offered a promise of a verdant planned community, with parks and lawns replacing the alleys and littered streets of the slums. But it, too, deteriorated as maintenance lagged. Apartments flooded, sewage backed up, and heating systems failed. When the Plan for Transformation got under way, Wentworth was given an extensive rehab. The units were modernized with new kitchens and bathrooms, the grounds were again landscaped, the parking lots expanded. Wrought iron fencing now ringed the perimeter.

All of that was good to Annie. But she also pointed out that the downstairs door to her building was broken. Even though it was locked, people just yanked it open. The buzzer in the apartment didn't work, either. She was assured that these problems would be fixed. She wanted to remain upbeat, even though with no laundry hookup she now had to drag a cart to a separate laundry building. And despite the demands she made while holding out in 1230 N. Burling, Ricks was relocated to an apartment with only three bedrooms. She slept in one of the rooms, Rose and her baby in another, and Reggie and his son stayed in the third. Raqkown, the youngest, was like Michael Evans from *Good Times*—the little fellow who slept on the couch. "Okay," she said. "We'll make this work."

She quickly learned that Wentworth was plopped down in the middle of a food desert. When a new White Sox stadium was built next to the old one, in 1991, it replaced much of the surrounding neighborhood, clearing homes as well as local businesses and the jobs they provided. The closest supermarket when she moved there was nearly two miles away. Most residents did their shopping at a gas station and liquor store on the south end of the development. One of Ricks's daughters sometimes drove her to a grocery store, but Annie liked to walk, so she also headed out on her own. She journeyed south a couple of miles to shop at Jewel or Walgreen's. She'd trek another half mile beyond that to visit an aunt in a senior center, and sometimes she went by foot to Oakwood Shores, a rehabbed public housing complex near the lakefront where some of her cousins lived. Ricks now cursed herself for not deciding where she wanted to go rather than letting the CHA pick. "We ended up there because my dumb butt didn't make them put in writing about a four-bedroom," she would say. "That was my mistake."

Annie didn't unpack entirely, as if she couldn't accept that

Wentworth was her home. In the living room she hung a photo from her wedding day. There were photos, too, of her deceased mother, who'd lived to ninety-four, and of Annie's children at their graduations and three pictures of President Obama. A couple of prayers and rules were taped to the wall—"Men don't wear hats in the house." But boxes were stacked unopened. Most of the trophies were never put on display. A bassinet was stuffed with clothes, and a grandfather clock remained taped shut from the day the Big "O" Movers carried it inside. One day Reggie took stock of their new apartment. "Tell you the truth, we downsized," he told his mother.

Annie Ricks couldn't deny it.

DOLORES WILSON

DOLORES WILSON DIDN'T return to 1230 N. Burling for the vigil. She was in her eighties then and couldn't stand out in the cold for hours looking at a lifeless wreck. She had dizzy spells, trips to the doctor, tests. "The longer you live, the house is going to tear down," she'd say. Plus, she was still familiarizing herself with the bus routes and the streets around Dearborn Homes. Most Sundays she did go back to Cabrini. Holy Family Church had a van that picked up its members who'd been scattered around the city, and for a small fee Dolores also took the Pace bus. She tried to make the 10:00 a.m. services and also the Bible classes that followed.

Her new neighbors liked to tell Dolores how bad the Dearborn Homes used to be. "They said the building I'm in was the worst," Dolores explained. "It was where the drugs were sold. The police wouldn't come in there, no security." When it opened, in 1950, Dearborn was the first public housing complex in Chicago with elevators—sixteen X-shaped buildings of no more than nine stories with a "towers in the park" design. In the 1940s and '50s, the CHA designated many of its new developments as "relocation projects," housing for the tens of thousands of families displaced by highway construction, slum clearance, and other large urban renewal efforts. Under the Plan for Transformation, Dearborn Homes became a new relocation project, undergoing a multimillion-dollar renovation. Thousands of families left public housing high-rises for a different rehabbed public housing site. "If I'm at Cabrini, and I choose Dearborn Homes, the public housing I'm moving into is nowhere near the public housing I'm moving out of," Lewis Jordan, the head of the CHA from 2007 to 2011, said. "People are exercising a choice: they know in the relocation process that they're going to 100 percent public housing that's 100 percent better."

"Some choice," Dolores would crack. When she moved into Dearborn, though, the floors were waxed and the elevators and walls clean. Trees and green grass surrounded the mid-rises. A security guard checked IDs in the lobby. A bus ran along State, trains were nearby, and the campus of the Illinois Institute of Technology sat just to the south. She did complain about the absence of any stores—the closest grocery then was fifteen blocks away—and about the size of her "teeny-weeny" apartment. Dolores would say of her place that she didn't have room to move around. "The elevator is in my bedroom, almost. I can sit in my living room and cook in the kitchen."

She liked that her daughter Cheryl lived next to her, on the very same floor, and Che Che visited often, and she sometimes looked after her great-grandson, who went to a school next door that she could see from her window. "All he had to do was take the elevator down and cross the street, and it's not even a traffic street," Dolores would say cheerfully. But the underenrolled school was one of the fifty shuttered by the city in 2013. From 2000 to 2010, Chicago's African American population fell by 181,000, a 17 percent drop, families deciding to quit the troubled city for the southern suburbs or the South. And eventually Cheryl had enough of Chicago and joined the exodus, moving to Atlanta to be closer to her own daughters. For all those reasons, Dolores couldn't muster the interest to do the little interior decorating that she'd always loved. She couldn't make her "way small" apartment feel like home. "I'm there," she'd say. "I'm surrounded by walls. But I'm just walking through."

Dolores said she no longer wanted to get involved in politics and activism, but she'd been saying that her entire adult life. She continued to obsess about prison reform and police abuse, about the number of unsolved murders in the city's black neighborhoods and all the money spent on building up the areas near downtown. She called her friends and wrote letters to local papers. She didn't attend the tenant council meetings at Dearborn—it was too far a walk across the development. But when her great-grandson's school was being closed, she went to protests to try to keep it open. "I could not imagine that they would just relocate children for the benefit of those with money or with political power behind them," Dolores said. And then she learned that the CHA hadn't fixed up or rented out almost 3,500 of its units, despite the fact that the federal government was still paying the agency whether the units were occupied or not. The CHA had also issued 13,500 fewer vouchers than HUD had funded. With tens of thousands of families on the waiting list for a subsidized rental, and hundreds of thousands more above the public housing threshold but struggling to pay rent on the open market, the CHA built up a reserve of $430 million, claiming it as some sort of rainy-day fund—while just about every other city and state agency looked to be in penury.

Dolores almost cried she was so mad. What about the rain that families in need of housing were feeling right now? She came to feel about a new mayor, Rahm Emanuel, the same way she did about the second Mayor Daley—"I hate their natural guts," she said. She joined a local group whose mission was to engage the Bronzeville neighborhood—her new neighborhood—in "civic capacity." "It sounds really elaborate, but it ain't," Dolores joked. When the group went to Springfield to protest even deeper cuts to social service programs, Dolores rode the bus downstate with her fellow demonstrators. As people marched around the state capitol, they shouted, "What do we want!" "When do we want it!" Dolores waved her fist and yelled the words, thinking at the same time that what she wanted was to sit down and to do it right now.

"Maybe after a while the protesting will pay off," she said with a sigh. "Maybe if we keep going, there will be some change."

DOLORES HAD IT better than many others who moved away from Cabrini-Green. They dealt with sketchy landlords, out-of-control utility bills, and an unfamiliarity with their surroundings that often put them in the way of danger. The gang structure in Chicago had fragmented into some 850 cliques, small groups of guys who grew up together, their collective identity organized around a corner or a park, the local school or the slaying of a friend. Their clashes with one another were mostly petty, personal, stirred up on social media, the kinds of conflicts that occur when every few blocks a different crew holds sway. Yet with more guns on the streets of Chicago than in New York City and Los Angeles combined, the arguments too often proved deadly. In 2012, Chicago saw 504 homicides, more than in any other city in the country and the most there since the early nineties.

Relocated public housing families were often blamed for the soaring rates of crime. There was an assumption, repeated by community leaders, the police, and homeowners, that the newcomers brought with them their "project behavior," a lack of socialization, as if they had re-created Cabrini-Greens wherever they settled. In South Shore, which took on more Section 8 renters than any other Chicago neighborhood, a resident described the arrival of high-rise families like a biblical plague: "It is as if the gates of Hell . . . opened and these people were let out. . . . I had to ask again, where did these people come from? And, lo, I was told they came from the projects, the CHA." It turned out the problem families he was describing were not from public housing, but no matter. Sometimes the behavior deemed "ghetto" wasn't the fault of the voucher holders but of their landlords. Families had to yell out their apartment windows to check on the door when buzzers and intercoms were broken and the requests to fix them went unheeded. A police commander who'd patrolled public housing high-rises before they were torn down said, "Ninety percent of the people in there were great people." It was about perception. "That stigma of the CHA," he added, "breeds the fear." Experts talked of a tipping point at which "crime would explode" as more public housing families were relocated to an area. A criminologist from the University of Illinois at Chicago speculated that the dispersal of CHA families had already led to a sharp jump in crime in the city's southern and western suburbs. And people in middle- and working-class Chatham, on Chicago's South Side, blamed the Plan for Transformation for their neighborhood's decline.

The communities with large numbers of Section 8 landlords, however, were already suffering in postindustrial Chicago. The Loop and the areas around it added 48,000 residents from 2000 to 2010, more than in any other city center in the country. Many black communities, on the other hand, continued to be drained of population, and they became even poorer and more perilous. Then the foreclosure crisis hit. By 2013, Chicago had 62,000 vacant properties, with another 80,000 foreclosures working their way through courts countywide. Two-thirds of the vacant homes were clustered as if to form a sinkhole in just a few black and Latino neighborhoods. One of every five apartments in South Shore was caught up in a foreclosure, along with thousands of single-family homes. Vacant houses were more likely to be used by bored kids, gangbangers, and drug users. Maps of the city's homicides, foreclosures, and public housing relocations looked almost like perfect overlays of one another. Families from Cabrini-Green and other high-rise developments didn't spread crime like an infectious disease; they were moved into tense areas far from the city center where the poor were already competing over shrinking resources.

That also meant relocated families were bound to exacerbate the problems. It would have taken uncommon coordination for the CHA, the police, and the schools to work in tandem on the relocations to avert conflicts. But that's exactly what was required. "It's like they took all the gangs and mixed them up," said the uncle of a sixteen-year-old who was shot to death in Englewood after he was moved from Robert Taylor. "Every project they shut down, they don't check where they put you. They just put you."

Eric Davis, the Cabrini-Green police officer who was part of the Slick Boys, lived in South Shore, and he now saw people all the time in his neighborhood who'd moved from public housing. They'd greet him as 21, and they'd laugh about something in the past that probably wasn't funny when it happened. Davis had retired from the force in 2007, after twenty years on the job, and he refused to set foot ever again on the land at Cabrini-Green. He'd show up to meet someone and linger just beyond the periphery of Seward Park, as if repelled by an invisible electric fence. He figured that he and every other police officer, resident, and activist who spent time at Cabrini suffered in some way from post-traumatic stress. His wife would catch him staring into space. "You in Cabrini, baby?" she'd ask, touching his arm. He had another reason never to go back—he wanted to preserve the purity of his final day on the job.

During the last two hours of his very last shift, Davis was called to 1017 N. Larrabee, J. R.'s old high-rise. A twenty-five-year-old woman was going into labor. The paramedics told her it was false labor and she had nothing to worry about. Davis had known the woman since she was a child, and she demanded to speak to him and only to him. It was her third baby, she said. By now she understood when a baby was coming, and it was coming. Davis agreed to take her to the hospital. The elevator was broken, and he held her arm as they walked slowly down each flight of stairs. They made it as far as the lobby.

"Twenty-One, I'm having my baby," she said. He got it, he told her, that's why he was taking her to the hospital. "No," she repeated. "Right now." She lay on her back on the floor, slid off her stretchy maternity pants, and told Davis to look. Bracing himself, he kneeled between her legs and saw the melon of the baby's head pushing its way out. Other residents gathered around. By the time Davis got the courage to reach for the emerging form, a hundred spectators had filled the lobby.

"Damn, I didn't know you guys delivered babies," an onlooker said. The woman spread her legs wider, and the rest of the baby started to appear—a face, shoulders, wriggling arms. Davis's hands were down there, wet to the elbows. Then, suddenly, miraculously, he was holding a tiny boy. He placed the baby in the mother's arms. Davis had arrested the child's father before, catching him in possession of an unlicensed gun and sending him to the penitentiary. But when the father showed up that night he embraced Davis like a brother. The mother said she wanted to name the baby Eric, after Davis. It was their first boy, though, and they gave him his father's name. But everyone called the child 21.

ANNIE RICKS

ANNIE RICKS BOASTED that she could adjust to anything. She'd seen it all. But when violence broke out at Wentworth Gardens, the longtime residents blamed the newcomers, the outsiders from Cabrini-Green. The Ricks children walked the grounds, and older tenants sucked their teeth, saying they didn't have problems before *they* showed up. An elderly woman announced for Rose Ricks to hear that she didn't like the Cabrini people and wanted them gone. A middle-aged man told Reggie, "I'm going to make you move." Annie's children were basketball players, and when Reggie and Raqkown tried to play on the courts beside their apartment, the guys there threw elbows and fists, trying to turn the game into a brawl. A group of boys jumped Reggie in the courtyard. The entire area seemed to stick together, so the odds looked always to be about fifty to one. Reggie had a gun held to his face, and his girlfriend who was six months pregnant was knocked to the ground and had to go to the hospital. J. R. Fleming's daughter with Donna had a pistol pulled on her as well. "There have been beatings. Children have had their wrists broken, jaw broken," J. R. told the *New York Times* about the Cabrini families who moved to Wentworth. "It's a powder keg."

The Ricks children came to appreciate all the more their mother's stubborn refusal to leave Cabrini-Green. Even amid the violence of their old home, they knew they could rely on allies there earned over a lifetime. It wasn't like that at Wentworth Gardens. "My sons didn't come from Cabrini-Green; they came from their mother's womb," Annie would say. "My sons don't sell drugs. They don't keep guns in the house. They're not in any gang. So you can't judge them. You know how people say you come from a different development. It's like a gang thing. People at Wentworth think you're stepping on their turf. This is not your turf. This is CHA's turf. You can't run me from my home. Because I do pay rent."

On an airless Saturday in the summer of 2012, Ricks went outside around midnight. She'd been indoors with her grandchildren for hours, upset with Rose for leaving the babies with her all day. She set up a small table, placing atop it a transistor radio and a can of bug spray. She popped open a bottle of iced tea she'd bought at the nearby gas station and tuned the radio to a gospel station. She had about ten minutes of relaxation before she heard screaming and cursing in the distance. She recognized the voice before she saw him—it was Reggie. He came sprinting toward her out of the darkness with Raqkown beside him and a mob of twenty men at their heels. Her sons had been at the apartment of a friend, another former 1230 N. Burling resident who'd moved to Wentworth. A group of guys had forced their way into the unit. Reggie and Raqkown managed to punch and push their way out before taking off across the interconnected courtyards. While Reggie was running and fighting, an elderly woman came outside to smack him with a broom; a man hit him over the head with the lid of a Weber grill.

Reggie and Raqkown dashed past their mother and up the stairs to their apartment, their pursuers rushing past her as well. Rose, who was inside, opened the door for her brothers, and the Wentworth guys charged in behind them, colliding into her. Annie Ricks ran up the stairs after them. "Home invasion!" she cried. In the cramped apartment, one of the invaders knocked a television off the wall and stomped his foot through it. They pushed over a chest of drawers and threw chairs. Reggie picked up a cooking pot off the stove and swung it. Rose armed herself with a mop. Their mother always carried a fist of keys with her, and she now punched with it. She had the aerosol can of bug repellant as well, and she sprayed it into any face close by.

"I'm just going to say it like this," Ricks said later that night, "we did whatever we had to do to get their asses out of our house. God told me to get the bug spray. I had to protect my sons. But I'm not mad. If I get mad, it wouldn't be me."

When the police arrived, they were what Ricks described as "Bridgeport police," white cops, seemingly from the traditionally Irish neighborhood just to the west. The officers laughed at Reggie, who was bleeding from his head. Ricks later asked for a police report, but the cops said only Reggie could request one. By that time, he was on his way to the hospital.

For two weeks after that night, Ricks stayed with a relative in the Cabrini rowhouses. Her daughter Earnestine put up Reggie and Raqkown. Rose moved temporarily to a place in the western suburbs. "It isn't safe for them at Wentworth," Ricks said. She complained about Wentworth Gardens to anyone who would listen, talking to Carol Steele and calling on the many reporters who'd interviewed her when she'd moved in the first place. "It's been hell," she insisted. "I'm going to say it just like this. They damn near killed Reggie. Ever since December 9, 2010, I can't get a good night's sleep." That was the day she left Cabrini-Green.

A social service provider asked Ricks if she wanted to go to a shelter. She didn't. A shelter wasn't any place to live. "That's like giving up," Ricks said. The CHA said she could move to the Cabrini rowhouses as part of the agency's victim assistance program. But she didn't want the rowhouses, either. "Why should I go from harm to harm, from hell's danger back to hell's danger?" Plus, the rowhouses weren't sized right for her family. With the home invasion, her old stubbornness returned. No way was she going to be cheated out of a four-bedroom again. Not this time.

The People's Public Housing Authority

J. R. FLEMING

IN THE MIDDLE of a 100 degree Wednesday, the same summer week in 2012 as Annie Ricks's home invasion, J. R. Fleming strode up to a brick bungalow on a South Side street lined with them. A man in an electric wheelchair whirred along the opposite sidewalk. A shirtless boy lay splayed like a cat across the hood of a shaded car. The neighborhood was known as Princeton Park, for the nearby planned community where Harvard and Princeton Avenues formed a loop around an eighty-acre greenway. The townhome subdivision had been built in the 1940s specifically for working-class African Americans. Black families soon bought up the bungalows on the surrounding streets, as blockbusting Realtors convinced white homeowners to sell low and flee to the suburbs. In the fifties, with the area's racial reconstitution nearly complete, Chicago aldermen agreed to build new public housing there—the Frank O. Lowden Homes, named for the Illinois governor who sent the state militia into the city during the 1919 race riot. The yellow-bricked house that J. R. now entered had been vacant for almost two years. The Anti-Eviction Campaign, the group he cofounded, had already broken in and taken over the property. Campaign members were in the living room stuffing the belongings of the last tenants into trash bags. "In an eviction," J. R. noted, "you took what you could, not what you should."

When J. R. emerged from the bungalow, a passerby demanded to know what they were doing on his block. A man in his fifties, he was a security guard at a lakefront condo tower several miles away. He said that the streets around him were pockmarked with abandoned houses, and he and the other homeowners who'd stuck it out had been calling the police to get trespassers out of the empty properties. J. R. greeted the news with a broad smile, as if he'd heard that they were cousins. He told the security guard that the people of Princeton Park weren't alone in their predicament. He said the foreclosure crisis had left Chicago gutted with abandoned properties—tens of thousands of them citywide. At the same time, the city had a shortage of 120,000 units of affordable housing and another 100,000 people sleeping in shelters or on streets or without a permanent place to stay. "Do the math," he demanded. "This country doesn't have a housing crisis. It has a moral and political crisis. Can you feel me?" J. R. denounced a housing industry whose greed brought low not only black neighborhoods but also the entire national economy. "The banksters are gangsters!" he yelled.

J. R. was enjoying his indignation. He lived for these public proclamations. Thirty-nine and still broad across the shoulders and chest, he had softened around the middle, and a skunk-like streak of gray ran up the center of his dreadlocks. He wore a medallion in the shape of the African continent over a loose-fitting Anti-Eviction Campaign T-shirt, which advertised that his organization had chapters not only in Chicago but also in Cape Town, South Africa. In salesman mode, J. R. explained to the man that his group had freed up a dozen abandoned homes in Chicago over the past year. Without seeking permission from any bank or the city, they'd pulled the boards off the doors and windows, made the properties livable, and moved in families in need of housing. "What we're doing is playing matchmaker," he said, laughing. "We're connecting home-less people with people-less homes." It was simple, logical, right? He said that the families they resettled into these properties wouldn't pay rent or a mortgage; many of them were former public housing residents and couldn't afford to. He chuckled deeply, asking the security guard if that wasn't preferable to a vacant home further deteriorating, turning into a haven for gangbangers or drug users.

The older man had been nodding in agreement, unable to interject. Finally, he said, "I'm glad to see you guys doing something positive." It did make sense to him. He was like a lot of people in neighborhoods such as this one: after living through decades of decline followed by the foreclosure crisis, they'd revised their economic understanding of real estate. Of course, they had considered their homes to be commercial assets, rock-solid investments that would steadily increase in value. But that was before their neighbors left and before the bubble popped. Black wealth in the country, already a fraction of that of whites', had been cut in half in the previous few years. Suddenly, a productive squatter next door seemed less like a criminal or a freeloader than a possible boon to a struggling community. The man did have one question, though. "How might I volunteer on that?"

The Anti-Eviction Campaign hosted its weekly Thursday night meetings at its headquarters, a corner lounge on the South Side that itself had been in foreclosure for a few years. The group's secretary was a woman named Shirley Henderson who had worked in the mortgage business herself. The financial crisis led to her losing first her job and then her home, from which she'd been illegally evicted, officers barging in without a warrant, guns drawn on her and her grandchildren. A year into her battle with a bank, she was at the courthouse when an officer told her about a boisterous young man from Cabrini-Green who'd been accompanying people to court. "When you're in foreclosure," Henderson said, "you look for help anywhere you can get it."

She now tried to sort J. R.'s roving weeks into some semblance of a schedule. There were days he led protests outside the downtown offices of Fannie Mae and Freddie Mac. Other times he enlisted campaign members to inundate banks with emails and phone calls demanding that a customer's delinquent mortgage be reworked. He led blockades outside homes to thwart banks from carrying out evictions.

He'd marched into a cramped Harris Bank with a team of protestors and a nonagenarian named Emma Harris who'd been set up by the lender with a subprime refinance with rates that eventually ballooned. After she went into foreclosure, the bank, rather than giving her a chance to rework her mortgage, "quit claimed" the property, forfeiting its ownership and handing the buildings over for free not to her but to a third-party developer. At the branch, J. R. thundered into a bullhorn, "This is a Harris versus Harris showdown!"

Back in 2003, J. R.'s awakening to public housing activism was sudden, the result of a police officer's fist to his face. How he came to guerrilla-style housing activism was more gradual. In 2009, J. R. was organizing against Chicago's bid to host the 2016 summer Olympics. No longer hog butcher or toolmaker, Chicago had by then come out of its postindustrial decline as a New Economy tourist destination for the world, a player with derivatives and trade shows, a city of big transportation hubs. While these industries breathed life into the city's central area, they couldn't support the sprawling Chicago that was. Chicago felt increasingly like a ring city, like London, Hong Kong, or Paris, with the moneyed elite residing in a prosperous core and poor people and minorities relegated to the resource-starved peripheries, the *banlieues*. And now Chicago was vying to build an aquatic center and a velodrome? It wasn't even housing its citizens or funding its schools. When Rio, not Chicago, was awarded the 2016 games, Republicans in DC spun it as a rejection of their caricature of President Obama's mobbed-up hometown ties and of his entire big-government liberal agenda. Locally, though, J. R. felt he'd helped avert a gross injustice. It was proof to him in the power of protest.

Later that year, J. R. learned that the United Nations special rapporteur on the right to adequate housing was carrying out an investigation across America. When she came to Chicago, J. R. was the one to show her around Cabrini-Green. The United Nations concluded that affordable housing conditions in the United States matched those in the third world. J. R. read the report and felt like he did in high school after his football team ran all over a rival. "I'm thinking, touchdown!" he said.

He believed that the international censure would shame Mayor Daley into preserving more of the city's public housing. He expected that the black president from Chicago would feel the scorn of the world and create an effective affordable housing policy for the nation. He thought that organizers in Chicago had finally saved Cabrini-Green. Of course, none of that happened. Despite the admonishment of the United Nations and J. R.'s own cries of wrongdoing, all the towers that remained at Cabrini-Green were torn down.

Even the remaining sliver of the Cabrini rowhouses, the last remnant of the old development, now looked to be doomed. The nearly 600 rowhouses were slated to be restored under the Plan for Transformation, counted among the 25,000 public housing units that the city promised to build or rehab. More than a decade into the plan, however, the city had created only 17,000 usable replacement units, and the number of public housing units at Cabrini-Green had been reduced from 3,600 to fewer than 500. Nevertheless, the agency now decided that even a single strip of rowhomes would be bad for both the neighborhood and anyone who had to live there in concentrated poverty. It didn't matter that the residents were surrounded by affluence or that the CHA was fine with relocating families to Wentworth Gardens, Dearborn Homes, Altgeld Gardens, or other public housing developments of concentrated poverty in less desirable real estate markets.

J. R. had to admit he'd been beaten and that the game wasn't even close. "I was piss fucking mad. I was bubbling," he would say. "Arguing with elected officials and the CHA doesn't matter. None of that shit matters. Because at the end of the day, they're going to do what they're going to do and we can't stop them from tearing down public housing." It was around then that a doctoral candidate in history at the University of Chicago named Toussaint Losier introduced J. R. to a South African housing activist. Losier had studied the direct-action tactics of the Western Cape Anti-Eviction Campaign, and he was showing the group's chairperson, Ashraf Cassiem, around Cabrini-Green. Cassiem listened to J. R. rant about the humanitarian crisis created by America's for-profit housing system. The South African laughed at J. R. for all his bluster: "The question is what are *you* willing to do when the government won't provide the necessary housing?" J. R. and Toussaint decided together to start a Chicago chapter of the Anti-Eviction Campaign. And unlike in South Africa, they wouldn't have to build lean-tos and shanties on government land; in the black neighborhoods of Chicago, they had all the empty homes they required.

One evening in the fall of 2012, J. R. helped lead a fund-raiser at a Bronzeville YMCA. For a modest donation, thirty people crowded into the basement to watch the film *Inside Job*, the 2010 documentary that explained the arcane origins of the housing crisis. Because many of them had been subject to subprime mortgages and foreclosure, the audience groaned and hissed as Matt Damon, the movie's narrator, described the proliferation of loans that by design were likely to lead to default. "That's their attitude to us," a man in the audience yelled. When Larry Summers, who had been one of President Obama's chief economic advisers, was shown as complicit in the undercutting of the country's financial oversight, the systems that might have reigned in the private sector's financial trickery, an elderly woman who was taking shaky notes on a palm-size pad let out a gasp. "Didn't Obama take him?"

For the people at the South Side YMCA, the movie explained that their fates had been determined by the actions of Goldman Sachs and Lehman Brothers, by the deregulation of banks in Iceland, by collaterized debt obligations and credit default swaps. Realtors, lenders, and rating agencies were all in cahoots. The whole vast system was stacked against them. US banks were two and a half times more likely to steer African Americans toward risky subprime mortgages than they were whites, even when black borrowers qualified for traditional, more secure loans. Nationally, black and Latino borrowers lost their homes to foreclosure at twice the rate of whites. And those around the room hadn't only lost their homes. Their neighborhoods were in ruins. After rushing to force out families in foreclosure, banks had failed to properly market, maintain, and secure the vacant homes, as the law required them to do. The homes turned into hulls, further depressing the value of occupied buildings around them. By 2012, 40 percent of the people who owned houses in Chicago's black communities owed more on their mortgages than their homes were worth.

When *Inside Job* ended and the credits rolled, J. R. stood in front of the group, bouncing on his toes, as if he'd just watched *Rocky*. He didn't find the movie depressing, he said. It inspired him. He'd seen it nineteen times and hoped to see it 150 more. It was true that Obama and his people couldn't be trusted. The president's men believed in housing not as a human necessity but as an engine of economic growth. "Yeah, we got *change*. Chump change. *Hope* don't give you housing," J. R. said, mocking the president's famous campaign slogans. "The market failed us. Harvard, Yale, and the University of Chicago failed us. Our government—*the* government—doesn't belong to us. Forget them, they forgot us." That's why they were now doing themselves what Franklin Roosevelt had done for the country in the 1930s. "If the government won't provide public housing for the people, the people must provide it for themselves," he announced. The takeovers of abandoned properties were infinitesimal compared with the demand, but it was a start. "What we need is the people's public housing authority."

UNTIL HE WAS eight, Rahm Emanuel lived in Uptown, the North Side Chicago neighborhood where his father settled not long after emigrating from Israel. But Emanuel's formative years were spent outside the city, in the wealthy North Shore suburb of Wilmette. Much has been made of the way Emanuel's parents trained him and his two brothers to spar verbally and come at life aggressively. His father, a pediatrician and former member of an Israeli paramilitary group, tested his sons' aptitude at the dinner table. Their mother worked on them culturally. Rahm and his brothers were sent to summer camp in Israel, and he served as a civilian volunteer in the Israel Defense Forces. The oldest brother, Ezekiel, went on to become a renowned oncologist and bioethicist; Ari, the youngest, with an infamous temper, rose through the ranks of Hollywood agents, taking over at the powerful William Morris Endeavor agency. Rahm, the middle child, excelled at ballet, earning a scholarship to study at Chicago's premier company, but he turned eventually to politics. He raised money—lots of it—for campaigns. Argumentative and profane, he relished the combat in every ask, demanding tens of thousands of dollars from donors who had the gall to suggest a pledge of only a few thousand. He worked as the chief fund-raiser for Richard M. Daley's first successful mayoral run, in 1989, and at thirty-two served as the national finance director for Bill Clinton's 1992 presidential campaign, before assuming a role as one of the president's chief advisers. When Emanuel returned to Chicago in 1999, it was for a job at the local office of a major investment-banking firm. In little more than two years with the company, with his long list of contacts in the Daley and Clinton administrations, he managed to earn $18 million. He spent $450,000 of it as he won a seat for himself in Congress in 2002. After Barack Obama tapped Emanuel to serve as his first chief of staff, critics characterized their shared Chicago roots as everything from provincial ineptitude to machine cronyism. When the administration pushed through its signature reform of the health care system, in March 2010, Rahm's backroom arm-twisting was hailed as a triumph of the "Chicago way."

Daley had been mayor of Chicago all that time, through most of the first Bush presidency, both of Bill Clinton's terms, the entirety of the George W. Bush White House, and into the Obama years. In 2011, having already surpassed the more than twenty-one years his father had held the office, Daley decided not to seek a record seventh term. Although he had won his last election, in 2007, with three-quarters of the vote, Daley had since seen his approval ratings fall to an all-time low. There was the city's failed bid to land the 2016 Olympic games as well as a fiasco involving the privatization of the city's parking meters. And maybe, too, he'd just had enough. The city's finances were in far worse shape than the mayor let on, and his wife of thirty-nine years, Maggie, was dying of breast cancer. With Daley's blessing, Emanuel came back to the city to run for mayor. After a Daley in charge for forty-three of the previous fifty-five years, Emanuel represented both a break from and a continuation of the past; he was something new, yet also a political and business insider, the type of pugnacious leader Chicagoans were conditioned to expect.

At Cabrini-Green, in her rowhouse offices, Carol Steele talked about her high hopes for Mayor Emanuel. Every politician and public official she'd known had looked on as public housing communities were wiped out, as poor people and their problems were forced out of sight. But with Emanuel she was willing to suspend disbelief. She had faith in Rahm not because of his ties to Obama. "Don't get me started on *that* one," she said. Emanuel made her think of the biblical story of Saul undergoing a conversion, transforming from a persecutor of Christians and into Paul, one of Jesus' apostles. "I think Rahm might be Saul going to Paul," she said. "I *hope* Rahm is Paul."

As mayor, Emanuel did throw himself into the idea of Chicago. "Chicago is the great American city, the most American of the American cities," he said. "New York City looks out onto the world. LA looks in a mirror. Chicago—it's where people from around the world and the country come to make a home." Emanuel had served on the board of the CHA when the Plan for Transformation was developed more than a decade earlier. During his bid for mayor, he'd said little about the unfulfilled plan or about public or affordable housing. Then in 2012, with Emanuel announcing that Chicago faced a budget deficit of $636 million and with the foreclosure crisis showing no signs of abating, his administration announced the creation of a Plan for Transformation 2.0, a "recalibration" of the original goals.

The new plan was much like the old plan. The CHA remained thousands short of reaching its promised goal of 15,000 new or renovated family units of public housing. With no sale of market-rate apartments to fund the building of low-income rentals, the city continued to add at most a couple hundred new public housing units each year. As part of the recalibration, the city did clarify that it wouldn't restore either the Cabrini rowhouses or the Lathrop Homes—both amid gentrifying neighborhoods on the North Side. And yet the demand for low-income housing in Chicago was as great as it had been in eighty years. When the CHA opened up a lottery in 2014, allowing people merely to sign up for an opportunity to get onto a public housing and Section 8 waiting list, 280,000 people applied, a quarter of all the households in Chicago. These, too, were people who'd come to Chicago to make a home.

J. R. FLEMING

THE VERY FIRST abandoned property the Chicago Anti-Eviction Campaign took over, in the summer of 2011, was on the 6700 block of South Prairie Avenue, in a section of the South Side known as Park Manor. The name of the century-old community had come to seem almost ironic, like a runty dog that people called Killer. There was one small park and no manors, just aging three-flats and single-family homes wedged into a crease formed by the crisscrossing of rail lines and expressways. Blocks were gap-toothed by the blight of empty lots and boarded-up homes, many of them marked with red "X" signs, alerting firefighters that the precarious structures should be left to burn. Dolores Wilson had lived on Prairie Avenue before coming to Cabrini-Green in her twenties, and around the time J. R. was eyeing vacant properties there, a driver took her past her old home. As Dolores rode by the building, she kept silent, ashamed to claim it as her own. The front door was open, the windows smashed, their tattered curtains rippling in the wind. Most of the surrounding buildings and stores that she remembered were gone, replaced by tracts of weeds and tall grass, befitting an actual prairie.

A friend told J. R. about the redbrick Victorian in Park Manor. Deutsche Bank had foreclosed on it two years earlier, and the house's owner, for all her efforts, couldn't convince the bank to consider a mutually beneficial modification of her loan. She finally walked away from it, moving to Philadelphia. But her foreclosure ended up among two thousand temporarily halted when lawyers working for the bank admitted to illegally altering documents. With the foreclosure in legal limbo, J. R. saw an opportunity. "When the ownership is complicated," he announced with a playful grin, "then it's community property."

What the Anti-Eviction Campaign wanted to do was technically illegal. But J. R. liked to boast that he didn't concern himself with the law. The takeovers "weren't legally right, but morally right," he'd proclaim. He'd offer up the example of the Underground Railroad, an audacious act of theft. He reminded people that until 1967 it was illegal in many US states for black and white people to marry. "We're challenging amoral laws by breaking them."

One of the Anti-Eviction Campaign's board members was a thirty-four-year-old former Cabrini resident named Martha Biggs. She persuaded J. R. to move on the Prairie Avenue house. "This is it," she told him. "This is where we can make our statement about the human right to housing." Her interest in the home was also personal, since she hoped to live in it with her four children. Martha grew up in one of the white high-rises at Cabrini, one of eleven children. When she was eighteen, her mother died; at twenty, and with her own apartment in the rowhouses, she was evicted for drug possession. Like many residents kicked out of Chicago's public housing at the time, she moved herself into one of the thousands of units the CHA had failed to refill once they became vacant. When she was put out of that Cabrini unit, she squatted in another one. The utilities were still in the name of the previous tenant, and after Martha landed a job at a hot dog factory, she got a tax return of $4,000 and offered to pay the entire bill due, a total of $2,000. But the CHA wouldn't take her money. The agency didn't put Martha out, but neither could she officially reside there. When her building was finally shuttered, she moved with her children to an apartment on the West Side. Then one day a sheriff showed up to kick them out—the building had gone into foreclosure. Her four children were between the ages of one and twelve. They slept on the couches or living room floors of family members. They stayed in homeless shelters, but at the shelters there were bugs and thefts and a feeling that people had quit trying. More often they huddled for the night in a parked minivan, Martha finding a way to keep her children neat and ready for school.

Tall and powerfully built, with ropy biceps, Martha was more than ready to work on the vacant house on Prairie. Scavengers had broken in and ripped out the pipes, toilets, radiators, ceiling fans, and cabinetry. And it wasn't like they did it carefully. There were holes in the walls and ceilings. Martha and other volunteers got to work drywalling and tiling and replacing what was taken. They stripped old paint and laid down fresh coats. They repaired windows and walls.

Six weeks after the start of renovations, the Anti-Eviction Campaign held a news conference on the front lawn to announce the home takeover. Martha stood beside J. R. and three of her children. Occupy Wall Street was then taking place downtown, and J. R. had befriended some of the activists, most of them young and white and looking for ways to shift from symbolic displays of outrage in the city center to something tangible in the areas hit hardest by the excesses of the 1 percent. A group of them had come out to the press conference, and they stood on the front porch, chanting, "Fight, fight, fight! 'Cause housing is a human right!" J. R. spoke first into the microphones set up by local news affiliates, settling into a preacher's rhythm. "Because of the government's inability to provide an answer to the homeless crisis that is plaguing our country, because of the banks' unwillingness to help homeowners, we have taken it upon ourselves, as men, women of our communities, to take back control of our communities." Then Martha, less prone to public declarations, began.

"Hello, my name is Martha Biggs, and I'm from Cabrini-Green."

On his national television show, Tavis Smiley covered Martha as part of his "Poverty Tour." The *New York Times* ran an article about the Prairie Avenue Victorian titled "Foreclosed Home Is a Risky Move for Homeless Family." But Martha didn't think the move too much of a risk. She lived on the first floor with her children, whom she enrolled in neighborhood schools. Other members of the Anti-Eviction Campaign, including J. R., sometimes crashed on the second floor. J. R. and Martha had also canvassed the block ahead of time, asking neighbors how they felt about a family moving into one of the several empty homes on their street. The idea appealed to them. They lent Martha rakes and gave her chairs and a china cabinet. She, in turn, used her fix it skills to do minor repair jobs for them. The possibility always loomed that a sheriff or a representative of the bank would show up to evict her. Believing she would be compensated for the improvements she'd done to the house, Martha kept receipts, and after a year she estimated $9,000 in parts and labor. And as at Cabrini-Green, there were dozens of other empty properties in the surrounding blocks where she could move next.

"I've been through so much, really, I feel like I can live any-where," Martha said. "As for property, I came from nothing, I can leave with nothing. They say, 'Who are you?' I say, 'Martha Biggs.' They say, 'What's your address?' I say, 'Earth.'"

J. R. happened to be sitting on Martha's stoop in the fall of 2012 when a man from the city's Department of Buildings rolled up to mark an empty house across the street for demolition. J. R. had been eye-ing the property, researching its tax history and record of ownership, thinking the Anti-Eviction Campaign could do something with that orange-bricked two-flat. He jogged over to the city worker, shouting, "Uh-uh, that's not going to happen." J. R. told the man it was crazy that taxpayers were cleaning up the mess of corporate giants who'd gone unpunished for their misdeeds. The city had spent $14 million in 2012 tearing down 736 vacant buildings, including 270 abandoned homes that the police identified as shelters for gangs and other criminal activity, and Emanuel's administration had 1,400 more on its demolition list. J. R. talked about the violence in the city and the black flight that was emptying out neighborhoods like Park Manor. The South Side population plunged by another fifty thousand people over a five-year stretch. He said the city shouldn't demolish something that could be turned into an asset, a home. The city worker didn't disagree. With a long list of properties to visit that day, he decided to move on to the next one.

J. R. broke into the house not long after that. On a weekday morning, he slid open a window off the front porch, then, unable to unstick the front door, kicked it open from the inside. Other members of the Anti-Eviction Campaign were waiting for him outside, smoking cigarettes. Thomas Turner wore a bike helmet because at six feet four he regularly smacked into low-hanging pipes or ceilings during these maneuvers. Thomas pulled a drill from a black duffel bag and began replacing the lock on the front door. Martha got started securing the rest of the house, screwing shut the heavy wood windows on the first floor. J. R. flipped light switches, trying to find out whether the electricity worked. When a ceiling fan began to twirl, he sang out, "We've got power!"

By then J. R. had entered hundreds of abandoned homes, each one a variation on the same theme of despair. He'd stumbled upon drugs and whatever paraphernalia people needed to use or make them. He saw the gathered sheets and worn-down mattresses of trick houses, the carcasses of dogs and cats and rats and possums and raccoons. Walking around the hundred-year-old house on Prairie now, he documented the state in which they'd found it, unconsciously filling every silence, belting out an off-key "If I Had a Hammer." He snapped "before" photos of a gaping hole in the ceiling, the kitchen stripped bare of its appliances and cabinetry, a bathroom scavenged of everything but a seatless toilet, the plaster and studs blasted to pieces. "This is not even about selling stuff," J. R. brayed. "It's 'I'm

going to break up a bunch of shit cause I'm mad and I got to go.'"
Windows were shattered or missing altogether. The flotsam and jet-
sam from capsized lives blanketed the floors—old winter coats and
pants, soiled grocery store bags, a crusted gallon jug of Open Pit
barbecue sauce. Lying in the corner of the dining room was a water-
stained "My First Birthday" photograph of a boy in a grown-up's Chicago Bears jersey and wool cap. On a low table in the living room
rested a solitary Bible. "There's always a Bible," J. R. noted.

Neighbors dropped by during the day. None of them could recall the house's last legitimate tenants. Martha had run off what she
called "crackheads" who had pulled up in a U-Haul to strip the place,
though she figured they later parked around back where she couldn't
see them. The scavengers had pulled the tiles off the walls in the
kitchen and bathrooms. A widower in his sixties who lived alone in
a nearby apartment mentioned a shooting that had happened on the
corner over the weekend. "I got it on video," J. R. said of the immediate aftermath. "Eight shots to the back!" He announced that he was
leading an antiviolence rally later that week, and almost as soon as
he said it he leaped up from the porch and chased down two lanky
teens passing by to invite them. The guys nodded with confusion as
J. R. talked excitedly about how together they would reclaim the block.

An hour into the Prairie Avenue takeover, Thomas Turner, refus-
ing help, was still struggling to install the new lock on the front door.
"Work smart, not hard," Martha scoffed, as she lugged a window she'd
found tossed in a closet. Thomas had rehabbed an abandoned single-
family house a couple of blocks east, rebuilding the gutted bathrooms
and kitchen. He bought used windows and doors or retrofitted what he
could find, often carting home large parts on his bicycle. There were
seven people living there now, including himself. "The homeless people
love it," he said. Formerly homeless himself, formerly incarcerated, formerly addicted to drugs, Thomas had stumbled onto the Occupy Wall
Street encampment after his most recent jail stint and joined the growing movement. Through the downtown demonstrators, he hooked up

with the Anti-Eviction Campaign, though J. R. had placed him on probation from the group earlier in the year. Thomas relapsed during the protests of the Chicago NATO summit in May 2012: activists crashing at his place invited him to share their drugs, and he'd accepted. But he was now clean and doing amazing work, J. R. said, proof that the home takeovers not only provided desperately needed housing but also put unemployed and underemployed people to work, training them in the building trades, all while beginning to stop the slide of a neighborhood.

J. R. and Martha followed a maze of exposed wires from the kitchen down into the basement. No water tank down there, but no rats or roaches, either. Someone had jerry-built a bed of flattened cardboard boxes in the clammy recess beneath the basement stairs. Surrounding it were carton after empty carton of Newport cigarettes. Imitating the hawkers who prowled street corners trying to make a few bucks, J. R. shouted, "Loose squares!"

At some point over the last century, the house's second floor had been turned into a separate apartment. Its appliances were also gone, the bathroom, too, a demolition site. Thieves had hauled off a cast-iron radiator, and judging from the cracked rails on the banister and a collapsed step, had realized its heft and simply rolled it down the stairs. But the three bedrooms were largely undamaged. The unit's thick wooden doors were adorned with what appeared to be their original glass handles. In the living room, the afternoon light poured in through floor-to-ceiling windows. The room contained a decorative fireplace and arched entranceways. The hardwood floor, still appearing relatively new after a hundred years, glistened. "The downstairs always be *whoop whoop*, and the upstairs always looks nicer," J. R. said.

Marveling at it all, he had a thought that often came to mind when he began work on one of these abandoned homes. He remembered Michael Jordan winning those six championships during the 1990s, saying to Bob Costas in the postgame interviews, "This one is special." That's what J. R. told himself as he bagged up the trash at the Prairie Avenue home. This is the one. This one can make a difference.

The Chicago Neighborhood of the Future

THE LITTLE BRICK chapel, at the intersection of Clybourn and Larrabee, was built in 1901, an outpost of the American Protestant Epis-
copal Church in the industrial river community. In 1927, the Near
North Side was mostly Italian, and Saint Philip Benizi, the parish
church led by Father Luigi Giambastiani along Death Corner, bought
the building and rededicated it the San Marcello Mission. Decades
passed and Cabrini-Green's twenty-three towers rose up around the
chapel, the public housing population soaring to eighteen thousand
and the Italians long departed. In 1965, the Saint Benizi Parish
church was demolished, but the San Marcello Mission, in the shadows of several white William Green high-rises, continued on, with
only a few dozen parishioners and a sole Sunday mass. The mission
tried to serve the residents of the high-rises, offering job training and
drug treatment. In 1972, a priest asked William Walker, the Chicago
muralist, to paint the plain building. Walker covered the outside
entrance with figures of different races, their giant circular faces
overlapping like a Venn diagram and their hands joining in embrace.
Bordering what was painted to look like a huge stained glass window, Walker included the words "Why were they crucified" and a
list of those suffering: Jesus, Gandhi, Dr. King, Anne Frank, Emmett
Till, Kent State. He titled the mural *All of Mankind, Unity of the
Human Race*, and it reflected a hope for the close-quartered divi-

sions of Cabrini-Green, Lincoln Park, Old Town, and the Gold Coast. The archdiocese of Chicago shut down the mission in 1974, and the building was taken over by the Northside Stanger's Home Missionary Baptist Church.

Four decades later, the neighborhood had transformed again. The Cabrini towers were no more, and the church sat in the backyard of the new multistory Target. The heavily trafficked streets were repaved and bike-laned. Up the block, an REI and a Crate and Barrel superstore had opened, an upscale movie theater and shopping center, an Apple Store and businesses for body sculpting. Where the Ogden Avenue Bridge had stood, there was now a skydiving facility, people paying $69.95 for a few minutes in a wind tunnel to experience the sensation of free fall. In 2015, Northside Stranger put its prime parcel of land on the market, asking $1.7 million for the 5,200-square-foot lot. In anticipation of a sale, the church was given a fresh coat of paint, the faded mural celebrating racial harmony whitewashed entirely.

For years, developers referred to Cabrini-Green as the "hole in the donut," the one area in the thriving city center where builders dared not go. No more. "Cabrini-Green is the Chicago neighborhood of the future," a realty company wrote. Circling the Cabrini land were new condos and luxury towers with outdoor pools and spas. Next to where Dantrell Davis had lived, on Oak Street, townhomes with floor-to-ceiling windows sold before completion. Boxy Parkside mid-rises now lined both sides of Division Street. Cabrini-Green tenants had filed a lawsuit with the city in 2013 to reopen the 440 shuttered rowhouses as public housing units. The suit was settled in 2015, nineteen years after the first redevelopment plan for Cabrini-Green was proposed. The rowhouses would almost certainly be demolished. But public housing residents would be mixed into whatever replaced the buildings, filling 40 percent of the units. There was a good deal of city-owned Cabrini property that had yet to be developed—empty fields and concrete tracts still sat where many of the high-rises stood. Public housing units would also be sprinkled into the dense array of residential properties that was sure to come on the rest of the seventy acres.

One Near North Side developer argued that the name Cabrini-Green no longer be uttered. "It's 'North of Chicago Avenue,'" he insisted. "NoCA is what everyone should be calling it. The name is without the stigma of Cabrini-Green." Yet even Chicagoans drawn to a hot new housing market were loath to adopt a New York–style neologism. The *Tribune* editorial page appealed to its readers in 2015 to come up with a name for the former Cabrini-Green befitting local customs. Among the hundred-plus submissions were Cooley Park, Gautreaux Town, Gold Coast West, North Branch, Old Ogden, Severin, Newbrini, Montgomery, Brother Bill, and Seward Green. But by far the name suggested most was simply Cabrini. "And why not name the neighborhood after Mother Frances Cabrini?" the paper mused.

"When I go to church now, I can hardly recognize the neighborhood," Dolores Wilson said. "Condos, townhomes, wealth. It's not the same." Her church, Holy Family Lutheran, was still there but struggling amid the changes. Newcomers to the neighborhood flocked to places like Park Community, a multistory gospel-preaching nondenominational church built a couple of blocks away. Park Community was "committed to being in the city, for the city." But Dolores appreciated that Holy Family was there at all. "People didn't believe it would stand this long—being Lutheran and in Cabrini, too! But GOD IS GOOD ALL THE TIME," she wrote in a letter to the editors of several local newspapers on the church's fiftieth anniversary.

With the motto "Many Voices. One Near North," the Near North Unity Program was a new institution in the neighborhood that was also committed to the past. It was created at the start of the Plan for Transformation's second decade to join together the changing area's disparate populations—the remaining Cabrini-Green families, the newer homeowners and renters, the new businesses, and the old community groups. Abu Ansari came over from his Parkside apartment for a time and led the meetings. "To assuage my guilt," he said. Kelvin

Cannon sometimes attended, standing in the back, and so did Carol Steele, one of Marion Stamps's daughters, and Brother Jim.

The group's success in drawing out the neighborhood's different "stakeholders" was evident in the ways their many voices often clashed. During one monthly meeting, white property owners peppered the area's police commander with questions about the open-air drug sales they'd witnessed on Larrabee Street, not far from their condo building. They couldn't believe that in the revived community, on the very same block as the new police headquarters, dealers could set up shop outside a corner store, with buyers loitering there at all hours of the day. A "Cabrini-Green problem" was being allowed to return, and they demanded that a cruiser be stationed at the intersection. Finally, a man who grew up in one of the Reds broke the protocol of raised hands and no interruptions. "It's loosies!" he shouted. "They're selling cigarettes on the corner, not drugs." It didn't make sense for someone buying drugs to linger. "You live in Cabrini-Green now," he said. "In the good end."

The Near North Unity Program led race and culture workshops for its members, and it evolved into one of the chief arbiters of the community's needs. The group set up a pen pal program among the fifth graders in the eight area schools, spread news of job openings and internships, organized hunger walks, and ran back-to-school fairs and neighborhood cleanups. It inaugurated a series of summer jazz concerts in the redesigned Seward Park. Anything to create "positive loitering" and "a new vision on Division," its leaders said. It became such a presence that developers now sought the group's support on proposed condo towers and revised plans for the Cabrini rowhouses. Jesse White brought architects out to a monthly gathering to talk about the designs for his new Jesse White Community Center, the thirty-thousand-square-foot facility built at a cost of $13 million on Chicago Avenue.

The Near North Unity Program also joined the fight to save Manierre, the elementary school by the Evergreen Terrace apartments just north of Division Street. Like the other fifty-four schools that Mayor Emanuel's administration said would be closed in 2013, Manierre was underenrolled, and the minority students who did attend underperformed by most measures. Jenner, south of Division, once the most crowded school in all of Chicago, had been rebuilt as part of the Plan for Transformation, and the state-of-the-art building could seat as many as a thousand students. But with the towers knocked down, enrollment hovered around two hundred, and two-thirds of those students were former Cabrini families who no longer lived in the district and traveled long distances each day.

The city proposed a reallocation of resources, combining the students from both schools into the new Jenner. But the neighborhood objected, saying the Hatfield-McCoy conflict between the young people on either side of Division Street was real and endured. A group of Jenner girls responded to news of the possible merge by beating up a Manierre middle schooler. A Jenner boy posted a "hit list" on Facebook, implying that the nine Manierre students he'd identified would be shot. J. R. Fleming spoke at one of several public meetings to protest Manierre's closing, asking Mayor Emanuel if, in Israel, he would be willing to send his children to a Palestinian school. He distributed copies of the United Nations Convention on the Rights of the Child, indicating that the city council was a signatory. "I would rather kill the budget than kill a child," J. R. said. In May, the mayor's office relented: Manierre could stay open. It was one of only a handful of the condemned schools to win a reprieve.

In 2015, the Near North Unity Program turned its attention to Jenner and its ongoing underuse. The group suggested a merge not with another Cabrini-area school but with an elementary school less than a mile east, in one of the city's wealthiest districts. Ogden International suffered from the opposite problem as Jenner. The Gold Coast area surrounding Ogden had exploded with new residential development in recent years, causing drastic overloading at the school. If Jenner and Ogden were combined, kindergarten through fourth grades could be housed at one campus and fifth through eighth at another. Not a single white family who'd moved to the Cabrini neighborhood had enrolled a child in all-black Jenner. But for those with infants or children-to-be, the possibility that one of the best schools in the city would, in effect, come to them was a kind of inner-city dream. Ogden parents who showed up at meetings to support the consolidation said they'd read the literature on school integration, and it revealed that higher-performing, wealthier students didn't suffer academically in these mergers. They praised Jenner's new principal, Robert Croston, a young alum of Harvard University's School Leadership program. At Jenner, he'd initiated a campaign to improve daily attendance; he started a career day and family math nights. He was trying to reinforce a culture of success at the school by dubbing it the NEST, an acronym drawn from a school credo: "I am Neighborly. I stay Engaged. I am a Scholar. I use Teamwork." And a great many people from both schools talked also of the social justice aspect of the merge. Nearly a century earlier, Harvey Zorbaugh had written in *The Gold Coast and the Slum* of these polar opposite communities that were only blocks apart: "All the phenomena characteristic of the city are clearly segregated and appear in exaggerated form." Here, at last, was a chance to join together the extreme contrasts of the area, to level this imbalance. At a meeting to discuss a consolidation that would begin no sooner than September 2018, an Ogden parent said, "We've forgotten about taking care of other people's children."

There was, as to be expected, a group of Ogden parents who were vocal in their opposition to the proposal. They worried about practical hurdles, like transportation between the two campuses. But they also felt that the Cabrini-Green neighborhood had changed, just not enough. "As Ogden parents we have been given virtually no chance to protect what we have planned for our kids' future here," a parent posted on an online forum. Someone else wrote, "I am all about the social development and upliftment of underprivileged kids and families, but it cannot come at the cost of compromising educational and

behavioral and safety environment for all the other kids." Cabrini
families expressed their own concerns. Tara Stamps, a daughter of
Marion Stamps and a longtime Jenner teacher, showed up at one
of the meetings with several of her colleagues, all of them wearing
"Straight Outta da NEST" T-shirts. She worried that the consolidation would not be a union of equals but a way to push out poor and
black people. The neighborhood had already lost a high school and
three elementaries as part of Cabrini-Green's demolition. When the
rest of the cleared site was finally developed, a third of the new units
would be reserved for public housing families returning to their "native land." Would a school filled with Gold Coast students now be
closed to them? "I really want you to understand with a sensitivity
that Cabrini-Green didn't represent just buildings. Those were families. Those were communities," Stamps said. "The reason you have
scores of our young people coming back in treacherous weather is
because they are rooted to the land. They have a blood memory there.
Their grandparents and their aunts and their cousins and their favorite memories were there."

ANNIE RICKS

Annie RICKS ENJOYED herself at the jazz concerts that the Near
North Unity Program hosted in Seward Park a few Fridays each summer. "I saw my family and good friends. The music was good," she
said. She ducked into the field house to say hello to James Martin of
the Slick Boys, the police officer known as Eddie Murphy who still
worked a second job at the front desk; little children greeted him
as "Mr. Murphy" as they entered the building. She shopped at the
Target before heading home. Annie was living again at Wentworth
Gardens. She called the company managing the property, demanding
they repair the front door to her building and replace the lights in
the parking lot. "It's pitch dark out here. You all still playing with

my life as well as my kids' lives," she complained. Sitting by her kitchen window, she counted thirteen lights out. "It feels like I'm on house arrest," she'd protest. "If I go out there, someone might sneak up and hit me in the head with a bottle." Nothing was repaired for months, until a young man was shot in the courtyard, and then the CHA brought someone out to replace the bulbs.

Annie decided she now wanted to leave Wentworth for Archer Courts, a rehabbed public housing complex in the city's Chinatown. She liked the way the place looked. On the two seven-story high-rises there, the chain-link fencing along the open-air walkways had been removed and replaced with alternating clear and frosted glass panels. The ramps no longer felt like prisons; they were bright, colorful spaces where residents could look out at the Chicago skyline and at their children in the playground below. The architect who did the redesign, Peter Landon, had proposed preserving parts of Cabrini-Green as well, saving some of the red high-rises and surrounding them with townhomes. "That kind of building with infill around it could be interesting," Landon said. "But there was no political will. You couldn't come in with a proposal that was subtle. Oh well, maybe next time around."

Ricks took the bus to the CHA's downtown offices. An official told her she could get on a waiting list for Archer Courts. Then over the next weeks and months, she documented every time she phoned the CHA to ask for an update. Eventually, an official suggested she try instead for Oakwood Shores, along the lakefront, and Ricks liked that idea, too. She had family there. She asked for the transfer. Expecting to move within a week, or maybe the week after that, she packed boxes and stacked them in her living room. She called the property manager at Oakwood Shores, telling her to get ready for Annie Ricks. Then another month passed, and she was still at Wentworth Gardens. Several months after the 2012 home invasion, Joe Peery, from Cabrini-Green, hooked her up with what Ricks called a "sweet lawyer." They met at the supermarket across from the Parkside condos, Ricks handing all her paperwork to the attorney in a thick binder. "They think they playing with a fool," she told him.

She checked in with the lawyer every few days, asking how her transfer was progressing. With no move in sight, though, she started to doubt whether her attorney was all that sweet. "When am I supposed to be moving?" she asked during one of their many phone calls.

"When we confirm your rent situation, when all of that is taken care of, we'll go from there."

"That's what you all on now?"

"Just trying to get them to confirm that you're eligible to be on the wait list for a four-bedroom."

"So how long will that be?"

"I wish I knew. I think they'll respond to me. I'll bother them until they do. I think this is going to be taken care of. It's not going to be done quickly."

"It should be quickly, because you're my lawyer."

"I may be a lawyer; however, I'm not a magician."

She ended up firing him. He'd been emailing the CHA since July, and it was now September. "I'm not prejudiced," she'd say. "But if I'd have been white he'd have moved me the very same day. He doesn't have to live in Wentworth Gardens, in the ghetto, as they say."

Then she had the idea of moving out of the city altogether, to suburban Oak Park. She thought she could switch to a housing voucher and find a place there, maybe a single-family home. The public transportation and the schools were excellent. Two people at the CHA encouraged this move, though soon they stopped returning her calls. But she remained positive, certain that a move was imminent. "God is going to bless me with an apartment in Oak Park," she'd say. "It's not going to be public housing where I'm going to live." She just needed to stick it out and fight. "I got nothing but time."

As she waited, one of her grandchildren was named valedictorian of her elementary school. Her thirty-three-year-old son Erskine married a girl from Dolton, the suburb where J. R. Fleming had lived as a teenager that was now 90 percent black. Ricks danced at the wedding. Now the newlyweds were expecting their first child—Ricks's thirty-eighth grandchild. On August 1, 2013, Annie's family celebrated her fifty-seventh birthday, buying her flowers and balloons and a purse.

She was out walking not long after that, venturing from Wentworth Gardens, when a burning pain shot through her left foot. She continued on, but soon developed calluses on her heel where they'd never been before. Then she could barely put any weight on the foot, and the family took her to the emergency room. When a doctor said Ricks had diabetes, she didn't believe him. How did she suddenly have diabetes? But she was sent to the county hospital, and a doctor there amputated her toes. She worried what she would do if she couldn't walk long distances. How would she go shopping, visit family, keep sane? The recovery was supposed to take less than a week, but Annie got an infection at the hospital. She had trouble eating and lost weight, the sharp bones of her cheeks never more pronounced. Weeks at the hospital turned into months, and Annie started to think she no longer was herself.

Her family rallied around her. Children crowded into her hospital room during the day, drawing pictures for their grandmother. One of Annie's daughters painted her fingernails, and another slept in a chair alongside her mother each night. As her spirits and her health improved, Annie was moved to Northwestern Memorial, where she began to rehab her foot. After a couple of weeks, she was taken outside in a wheelchair for a stroll. She wasn't walking herself, but at least she was moving about, and Annie raised her arms triumphantly. She started talking again about where she would move after she was discharged in the coming days. "I need to laugh sometimes instead of cry," she said. "If I'm going to cry, it won't get me out of Wentworth Gardens."

But before the Northwestern doctors signed her release papers, Annie came down with pneumonia. She was relocated to another hospital, where the doctors said the scars around her toes had never healed properly. They intubated her to make sure she ate enough, but now she couldn't talk. She didn't smile. She needed help breathing. After so much time in bed, she had sores all over her body. Her daughters complained that the nurses there were ignoring their mother, failing to turn her properly. They pointed to the holes that were forming in their mother's back. Ricks's gaunt arms were black where months of IVs had ruptured her veins, and she whimpered through a tube.

A doctor came into the room one day to talk to the family about removing Annie from the feeding tube and the assisted-breathing apparatus. It took a while for the Ricks children to understand that he meant to let her die. He claimed she was unresponsive. They argued that their mother was heavily sedated. Each time she came off the assisted breathing, they said, she breathed on her own for longer stretches. She was the toughest person they knew. Their mother would never give up. The family hired a lawyer they heard about on television, a malpractice attorney who asked them repeatedly if there had been a bed fall, because that would be a "slam dunk." They told him, no, there hadn't been a fall, and he ignored their calls.

Then on November 16, 2014, with gospel playing and her family gathered around the hospital bed, the life passed from Annie Ricks. She was fifty-eight. "You walk in with a sore foot, and you never leave," a daughter said bitterly. The family struggled to pay for the funeral. One director simply walked out of the room when he heard they didn't have insurance to cover the cost. But they managed. They had a friend who could do the makeup; others chipped in for the coffin and the flowers and the food for the repast. The services were held at Kingdom Baptist Church, on the West Side. It had been four years since Annie Ricks left Cabrini-Green, and more than a hundred people from the old neighborhood showed up to pay their respects. At the memorial, her children talked about their mother's stubbornness, her determination to provide for them: "What she didn't have, she made sure we had it." In front of the pews, Reggie said he didn't care that he had a terrible singing voice, and he started in on an R. Kelly song. "Dear Mama, you wouldn't believe what I'm goin' through / But still I got my head up just like I promised you." Rose, crying, couldn't speak. Kenton, Ricks's fourth of nine sons, said they all learned from her example: "Be strong, take care of the kids, take care of family." She made them all better people. "She'd do anything for anybody," he said. "She was just love."

KELVIN CANNON

J. R. FLEMING attended Annie Ricks's funeral. "She was the last voice, the last resident's voice," he said. "She talked shit, too. She would tell it like it is." Kelvin Cannon was there as well. "Ms. Ricks fought for something she felt was right. She fought for her home," he said. "I admire her for that. If more people fought like her, maybe what happened to Cabrini would be different." After Cannon's time as tenant council president ended, he started looking for full-time work again. Despite what some thought, no one handed him a job as part of a quid pro quo. And he didn't run up to Jesse White, his old PE teacher, and ask to be hired. That's not how it was done. At almost eighty, White had been Illinois secretary of state since 1999 and Democratic committeeman of their ward for even longer, all while appearing at hundreds of Jesse White Tumblers performances each year. When Mr. White distributed food in the neighborhood, Cannon came out in the cold and hauled turkeys and hams off an eighteen-wheeler. Cannon appeared at fund-raisers and volunteered on campaigns, knocking on doors and passing out literature. He worked the back-to-school picnics. He hung out at every weekly ward night, too, at the Jesse White Tumblers' offices, on Sedgwick, beneath the Marshall Field Garden Apartments. Up until 2017, the horses that pulled carriages on the Gold Coast were stabled across the street, and the block smelled of hay and manure.

The ward offices also had the feel of a bygone era, as if White's own beloved mentor, George Dunne, were still doling out the favors and the first Richard Daley were reigning over the city. But it was White who presided behind a short desk, in a room the size of a closet, its walls lined with his memorabilia—campaign posters and photos of him on the Cubs and with the 101st Airborne. One after another, constituents were led into the small office, the door was shut, and the person asked humbly for assistance. A man wanted to turn an unused parcel of CHA land into a boarding house for military veterans. A woman was in foreclosure on her apartment and hoped Mr. White might talk the bank into giving her a little more time. Someone needed a suspended driver's license reinstated. If the person was from the neighborhood, White could name his or her uncle, grandparents, and cousins. "I saw your father run track," he'd tell a mother of three. Then he'd say, "Reach out to Annette in my office, she'll take care of it," or, "Call me tomorrow, here's my card, I'll see what I can do."

Boys always showed up, hoping for a spot on White's famous tumbling team. "I like smart people in my program. I don't like dull knives," he'd lecture the teen. "Keep your pants up, don't show your underwear. Don't say 'yeah,' say 'yes.'" He'd had 16,500 people flip and leap in the program since the 1950s and he said fewer than 150 of them had gotten into trouble with the law. And just when it seemed he would toss the dull knife of a teenager out on his sagging pants, he told him to be at the gym on Monday, writing the address on a slip of paper and making the young man repeat the directions to get there.

At one of these ward nights, White hired a couple of his regulars to do maintenance work for the state. The job required a third crew-member, and one of the men pointed to Cannon, who as usual was nearby. "Do you know Cannon?" the man asked. "Yeah, I know Mr. Cannon," White said. "I've known him all his life. I used to spank him." And that's how Cannon became an employee of the state of Illinois.

Cannon hoped to keep the job with the state until he got his pension in a few years, but he also had bigger dreams. He'd been asking Mr. White to help him get a pardon for his one felony conviction, a lifetime ago. He talked about going into law enforcement or opening up a restaurant or maybe trying his hand at politics. "Who would be better than me?" he'd say. "I know the politics and I know the people. I've run for stuff before. I'm streetwise and political-wise." Cannon understood that life was fickle, that it could take any sort of turn. He believed he had to be ready for whatever opportunity arose. Anything was possible. Your past didn't have to define you. "I've come a long way," he would reflect. "Cabrini-Green has come a long way."

J. R. FLEMING

ONE SUMMER NIGHT, J. R. and other members of the Anti-Eviction Campaign boarded a boat docked at the throat of the Chicago River. Brother Jim was raising money for Brothers and Sisters of Love, the Catholic ministry he started with Brother Bill before Bill retired, and he'd invited church leaders and benefactors to take a cruise with him and some of the Cabrini-Green people he'd worked with over the years. J. R. moved to the upper deck, standing on the prow as they floated through space and time. Jean Baptiste Point DuSable, a Haitian settler, had sailed up this river in the 1770s, setting up a trading post and becoming Chicago's first permanent resident. Slag and timber and livestock had followed, the river's direction reversed by 1900 to flush putrid waste downstate. Now the Trump Tower stood before J. R. like a giant switchblade mugging the sky. Tourists who were gathered on the bridges overhead called down greetings, and kayakers paddling alongside the boat waved in fellowship. The boat slipped by the old and the new, past stone buildings and glass spires, the clock tower of the Wrigley Building and the ramparts of the Merchandise Mart. The waterway and the reflective surfaces around it sparkled like treasure.

One of J. R.'s companions on the river was Raymond Richard, who'd grown up in the Cabrini high-rise known as the Castle. Richard had recently started a group called Brothers Standing Together that tried to help the formerly incarcerated find work and avoid a return to prison. He pointed to the berth beneath Wacker Drive where he'd slept for years when he was on his "tramp trail," a zombie strung out on heroin. "I lost all self-esteem and the will to live," he said. "My family would feed me through the door, because I would steal my mother's food stamps to buy drugs." He could see that people were living on Lower Wacker still.

Mayor Emanuel dismissed the notion that Chicago was divided into two separate cities of haves and have-nots, of the center and everywhere else. "Downtown versus the neighborhoods is a false dichotomy," he'd insist. But J. R. spent his days in that other Chicago of abandoned homes and joblessness, of black flight and the meanness born of despair. All around were struggling schools and shuttered businesses—half the young black men in Chicago were unemployed, more than in any other big city in the country. Of the thousands of shooting victims in the city each year, most were from these same neighborhoods. Even the "contract sales" from the middle of the last century, back when public housing looked like paradise, had returned. National investment firms were buying up distressed properties and selling the equity-less and exploitative contracts-for-deed on these homes to low-income buyers who likely couldn't secure a mortgage.

The boat sailed beside the corncob towers of Marina City. Charles Swibel had built them, and Curtis Mayfield had moved there from the Cabrini rowhouses. Then the boat turned right onto the river's North Fork, nearing Cabrini-Green, and J. R. and his friends talked excitedly about where along the banks they'd walked with girls or ran from police officers. They'd held jobs selling newspapers, picking up their supplies from the *Tribune* printing house there. They'd played games around the fortress of the Montgomery Ward warehouse. One of them still lived in the Cabrini rowhouses, almost visible now from the river. The plans to redevelop the rowhouses were moving along slowly, and Brother Jim was trying to help the resident's youngest brother—a teenager who'd been caught, this time, with a gun. Their mother wasn't giving up on her child, but she also didn't want to give up on herself, she said. She didn't want to lose her chance to stay in the neighborhood. She'd seen so many neighbors kicked out of public housing because of their children, and every time she turned around, her son was into something else. "I'm the opposite of optimistic," she'd told Jim.

J. R. was living on the far North Side. He'd met someone who loved him for his activism, and they had a son together and soon married. J. R. became a grandfather. He was still rehabbing vacant properties and trying to turn the abandoned homes into community assets. It seemed far-fetched and unfair to believe that the prosperity around the river would radiate out like a star and nourish those satellites distant in space and time. So J. R. was working with banks and nonprofits as well to establish a community land trust, in which everyone in an area pooled resources and collectively owned properties that couldn't be bought or sold for profit. He was now buying foreclosed homes, too, training young people to repair them and selling them below market value. "I want my legacy to reflect that I cared," he said. "My children need to know that their father came from Cabrini-Green, this great community, and he did something productive and positive."

The boat reversed course and began to drift back toward Lake Michigan. The sky was draped in darkness. A light rain started to fall. The common sentiment was that the people who lived in the luxury high-rises around them had created their own good fortune; the generous government benefits and tax breaks they received were rightfully earned. "That makes me smart," Donald Trump had said on his way to the presidency about paying no federal income tax. For those remaining in the Cabrini rowhouses or the Wild 100s or Englewood or Little Village or North Lawndale—there was no political will anymore for the government to step in and transform the blocks that had been left to wreck. No one cursed the city louder for its inequity and cruelty and racist history than J. R. But also no one cared for it more. "I am Cabrini-Green," he liked to proclaim. But Cabrini-Green was also Chicago, in all its ceaseless glory and failure. J. R. bellowed into the mist of the night air, "I love my home!"

Acknowledgments

First and foremost, I want to thank Dolores Wilson, Kelvin Cannon, the late Annie Ricks, and Willie J. R. Fleming for sharing so much about their own lives with me. I am indebted as well to hundreds of others who lived or worked at Cabrini-Green—this book exists because they sat for interviews, let me into their homes, met me in Seward Park, or talked on the phone for forty-five minutes or several hours. Carol Steele, the president of the Cabrini-Green tenant council for the seven years I worked on this book, welcomed me into her rowhouse office, and the members of the Near North Unity Program allowed me to take part in their meetings. The staff at the Chicago Housing Authority assisted with my research, and Keith Magee, Todd Palmer, and Lisa Lee supported my work during their tenures at the National Public Housing Museum.

This book would have been much harder to write without the many public housing experts who graciously answered my questions and shared their own insights: Marilyn Katz, Lawrence Vale, Cassie Fennell, Brad Hunt, Julia Stasch, Janet Smith, Sudhir Venkatesh, Jim Fogarty, Peter Landon, Eric Davis, and Susan Popkin. And I owe a special thank-you to Larry Bennett for lending me several bankers boxes' worth of his research and field notes. In Chicago, other writers, reporters, photographers, visual artists, and documentary filmmakers generously discussed their own work and directed me on my own: Natalie Moore, Alex Kotlowitz, Ronit Bezalel, Nate

Lanthrum, Jamie Kalven, Monica Davey, Ryan Flynn, Jan Tichy, and Megan Cottrell.

At *Harper's Magazine*, where I was an editor for a time, my former colleagues have edited my writing, offered guidance and support, and helped me professionally and personally in numerous ways. Among them, Jennifer Szalai first suggested I write about the demolition of the last Cabrini-Green high-rise. Christopher Cox, Rafil Kroll-Zaidi, Stacey Clarkson, and Alyssa Coppelman helped turn that idea into a magazine feature. Roger Hodge, Ted Ross, and Donovan Hohn gave tips on the book proposal, and I've relied on Bill Wasik's sage advice too many times to recall. At the *New York Times Magazine*, Joel Lovell masterfully helped shape what would become a piece of this book into a feature story.

I was lucky to have writer friends who listened to me over the past several years talk about my progress on this book, and lack of it, and who offered the needed commiseration and counsel: Rachel Kaadzi Ghansah, Claire Gutierrez, Ethan Michaeli, Micah Maidenberg, Matt Power (pouring libations for you *Cooley High*–style), Mark Binelli, Meg Rabinowitz, Paul Kramer, Will Howell, Yuval Taylor, Amanda Little. Maya Dukmasova and Bill Healy helped with research and insight. I am indebted to Gideon Lewis-Kraus for a major assist early in the process of this book and to DJ Pat Rosen for a couple near the end. I was fortunate to connect with Robert Gordon, who drew the incredible maps at the start of this book, and with the photographer Jon Lowenstein. A huge thank-you to Wells Tower and Audrey Petty who read portions of this book, and who were steady sources of support and inspiration. And an impossible-to-repay debt to Adam Ross, who was there with me every step of the way of this endeavor.

To Jonathan Jao, I am deeply thankful for his editing, encouragement, and guidance. I thank Sofia Groopman and the others at HarperCollins who shepherded this book to publication, and Tim Duggan for first taking a chance on it. Many people think they have the best literary agent; I really do in Chris Parris-Lamb, whose careful reading of my writing and passion for this book kept me believing it was worth completing.

The most heartfelt thanks go to my family. My two brothers from other mothers, fellow writers, insightful readers, and my constant companions in this thing called life: Khalil Muhammad and Sascha Penn. My brilliant brother, Jake; my parents, Ralph and Ernestine; Jacqueline Stewart, Maiya and Noble Austen, and Carol House. And the reasons I'm anything: Danielle, Lusia, and Jonah, who turn a house into a home.

Bibliography and Notes on Sources

In 2010, I began reporting an article for *Harper's Magazine* about the closing of the last of Cabrini-Green's twenty-three high-rises. I started working on this book soon after that. I spent hundreds of hours interviewing the people who appear in these pages. I spoke as well to numerous other Cabrini-Green residents, to officials from the Chicago Housing Authority and other city agencies, to local politicians, teachers, and principals, to social service providers, affordable housing advocates, community activists, lawyers, architects, police officers, clergy, business owners, developers, building managers, and residents from the surrounding area. My reporting took me to community meetings, public forums, court records, police reports, and people's homes. And I benefited greatly from the generosity of many journalists, researchers, academics, filmmakers, and photographers who were kind enough to speak with me about their own work and share their resources.

For the purposes of writing this book at least, the media's long fascination with Cabrini-Green and the Near North Side slum that preceded it proved tremendously useful. I relied on reporting that appeared in the *Chicago Tribune*, the *Chicago Sun-Times*, the *Chicago Defender*, the *Chicago Reporter*, the former *Chicago Daily News*, the *Chicago Reader*, *DNAinfo Chicago*, *Residents' Journal*, *Chicago* magazine, *Crain's Chicago*, the website Forgotten Chicago, and radio station WBEZ. For television news and other video footage featuring Cabrini-Green, I used the collections at the Museum of Broadcast Communications, Vanderbilt University's Television News Archive, and the Media Burn Archive. I found reports, correspondences, pamphlets, maps, and other historical documents in the archives of the Chicago Housing Authority, the records at the Metropolitan Planning Council, and in the holdings of the Chicago Public Library, the Chicago History Museum, the University of Chicago, and the University of Illinois at Chicago.

Here, more specifically, are the sources that were most instructive for each chapter.

CHAPTER ONE: PORTRAIT OF A CHICAGO SLUM

In writing this chapter, I relied on early Chicago Housing Authority pamphlets and publications, numerous accounts in the local Chicago press (some dating back to the nineteenth century), the notes and records of J. S. Fuerst (kindly lent to me by his daughter, Ruth Fuerst), and, of course, the many conversations I had with Dolores Wilson. The following published sources were especially useful:

Abbott, Edith. *The Tenements of Chicago, 1908–1935*. Chicago: University of Chicago Press, 1936.

Bowly, Devereux, Jr. *The Poorhouse: Subsidized Housing in Chicago, 1895–1976*. Carbondale: Southern Illinois University Press, 1978.

Chicago Housing Authority. *Cabrini Extension Area: Portrait of a Chicago Slum*. Chicago Housing Authority, 1951.

Cronon, William. *Nature's Metropolis: Chicago and the Great West*. New York: W. W. Norton, 1991.

Drake, St. Clair, and Horace R. Cayton. *Black Metropolis: A Study of Negro Life in a Northern City*. New York: Harcourt, Brace and Company, 1945.

Fuerst, J. S., and D. Bradford Hunt. *When Public Housing Was Paradise: Building Community in Chicago*. Westport, Conn.: Praeger, 2003.

Guglielmo, Thomas A. *White on Arrival: Italians, Race, Color, and Power in Chicago, 1890–1945*. New York: Oxford University Press, 2003.

Hunt, D. Bradford. *Blueprint for Disaster: The Unraveling of Chicago Public Housing*. Chicago: University of Chicago Press, 2009.

Meyerson, Martin, and Edward C. Banfield. *Politics, Planning, and the Public Interest: The Case of Public Housing in Chicago.* Glencoe, Ill.: Free Press, 1955.

Michaeli, Ethan. *The Defender: How the Legendary Black Newspaper Changed America: From the Age of the Pullman Porters to the Age of Obama.* Boston: Houghton Mifflin Harcourt, 2016.

Petty, Audrey. *High Rise Stories: Voices from Chicago Public Housing,* Voice of Witness. San Francisco: McSweeney's, 2013.

Philpott, Thomas Lee. *The Slum and the Ghetto: Immigrants, Blacks, and Reformers in Chicago, 1880–1930.* Belmont, Calif.: Wadsworth Pub. Co., 1991.

Vale, Lawrence J. *Purging the Poorest: Public Housing and the Design Politics of Twice-Cleared Communities.* Chicago: University of Chicago Press, 2013.

Wright, Richard. *12 Million Black Voices: A Folk History of the Negro in the United States.* New York: Viking Press, 1941.

Zorbaugh, Harvey Warren. *The Gold Coast and the Slum: A Sociological Study of Chicago's Near North Side.* Chicago: University of Chicago Press, 1929.

CHAPTER TWO: THE REDS AND THE WHITES

For this chapter, I used CHA publications from the 1940s, '50s, and '60s, including the agency's yearly statistical reports, along with hundreds of newspaper accounts from that era. Particularly helpful were the "North Side Observer" columns in the *Chicago Defender* written by Margaret Smith. I relied also on interviews with Dolores Wilson and other early residents of high-rise public housing in Chicago, and with Richard M. Daley and many public housing experts.

Art Institute of Chicago, et al. Chicago Architecture and Design, 1923–1993: *Reconfiguration of an American Metropolis.* Chicago: Art Institute of Chicago, 1993.

Black, Timuel D. *Bridges of Memory: Chicago's Second Generation of Black Migration.* Chicago: Northwestern University Press, 2007.

Bowly, Devereux, Jr. *The Poorhouse: Subsidized Housing in Chicago, 1895–1976.* Carbondale: Southern Illinois University Press, 1978.

Butler, Jerry, and Earl Smith. *Only the Strong Survive: Memoirs of a Soul Survivor.* Bloomington: Indiana University Press, 2000.

Cohen, Adam, and Elizabeth Taylor. *American Pharaoh: Mayor Richard J. Daley: His Battle for Chicago and the Nation.* Boston: Little, Brown, 2000.

Fuerst, J. S., and D. Bradford Hunt. *When Public Housing Was Paradise: Building Community in Chicago*. Westport, Conn.: Praeger, 2003.

Guglielmo, Thomas A. *White on Arrival: Italians, Race, Color, and Power in Chicago, 1890–1945*. New York: Oxford University Press, 2003.

Hirsch, Arnold R. *Making the Second Ghetto: Race and Housing in Chicago, 1940–1960*. Chicago: University of Chicago Press, 1998.

Hunt, D. Bradford. *Blueprint for Disaster: The Unraveling of Chicago Public Housing*. Chicago: University of Chicago Press, 2009.

Mayfield, Todd, and Travis Atria. *Traveling Soul: The Life of Curtis Mayfield*. Chicago: Chicago Review Press, 2016.

Meyerson, Martin, and Edward C. Banfield. *Politics, Planning, and the Public Interest: The Case of Public Housing in Chicago*. Glencoe, Ill.: Free Press, 1955.

Royko, Mike. *Boss: Richard J. Daley of Chicago*. New York: Dutton, 1971.

Vale, Lawrence J. *From the Puritans to the Projects: Public Housing and Public Neighbors*. Cambridge, Mass.: Harvard University Press, 2000.

———. *Purging the Poorest: Public Housing and the Design Politics of Twice-Cleared Communities*. Chicago: University of Chicago Press, 2013.

Werner, Craig Hansen. *Higher Ground: Stevie Wonder, Aretha Franklin, Curtis Mayfield, and the Rise and Fall of American Soul*. New York: Crown Publishers, 2004.

Whitaker, David T. *Cabrini-Green in Words and Pictures*. Chicago: W3, 2000.

CHAPTER THREE: CATCH-AS-CATCH-CAN

Kelvin Cannon and others who grew up in the William Green "Whites" in the 1960s and '70s told me of their adventures beneath the Ogden Avenue overpass and their fears of its fabled witch. The website Forgotten Chicago did a deep dive into the history and erasure of parts of Ogden Avenue. I spoke as well to Jesse White and several of his former tumblers and Boy Scouts, including Richard Blackmon and Perry Browley. Many Cabrini-Green residents and teachers from the neighborhood schools I interviewed shared their memories of the rioting after Martin Luther King Jr.'s assassination, but I also learned much from the recollections assembled in David Whitaker's book cited below. The *Defender* covered King's trips to Cabrini-Green and the school boycotts there. I benefited again from CHA historical records, and I need to mention the debt I owe to D. Bradford Hunt's magisterial history of the rise and fall of Chicago public housing, *Blueprint for Disaster,* which I cite below and throughout this bibliography.

Cohen, Adam, and Elizabeth Taylor. *American Pharaoh: Mayor Richard J. Daley: His Battle for Chicago and the Nation.* Boston: Little, Brown, 2000.

Hunt, D. Bradford. *Blueprint for Disaster: The Unraveling of Chicago Public Housing.* Chicago: University of Chicago Press, 2009.

Vale, Lawrence J. *Purging the Poorest: Public Housing and the Design Politics of Twice-Cleared Communities.* Chicago: University of Chicago Press, 2013.

Whitaker, David T. *Cabrini-Green in Words and Pictures.* Chicago: W3, 2000.

CHAPTER FOUR: WARRIORS

For this chapter I made use of the onslaught in coverage of Cabrini-Green after the murders there in 1970 of two police officers. Hundreds of stories appeared in the media, and the CHA documented carefully its initiatives to improve the now-infamous public housing development. Residents of Cabrini-Green shared with me their recollections of this pivotal moment in their experiences there. Kelvin Cannon's memories were invaluable for this chapter, as were interviews I did with Dolores Wilson, Jesse Jackson, Burt Natarus, and many others. A longtime alderman of the Cabrini-Green area, Natarus donated his papers to the University of Illinois at Chicago, and the archive helped me with this chapter and others.

Blackmon, Richard, Jr. *Pass those Cabrini Greens, Please!!! (With Hot Sauce).* Chicago: 714 Productions, Inc., 1994.

Cohen, Adam, and Elizabeth Taylor. *American Pharaoh: Mayor Richard J. Daley: His Battle for Chicago and the Nation.* Boston: Little, Brown, 2000.

Dawley, David. *A Nation of Lords: The Autobiography of the Vice Lords.* Garden City, N.Y.: Anchor Press, 1973.

Freidrichs, Chad, Brian Woodman, and Jaime Freidrichs. *The Pruitt-Igoe Myth.* DVD. [United States]: First Run Features, 2011.

Hagedorn, John, and Perry Macon. *People and Folks: Gangs, Crime and the Underclass in a Rustbelt City.* Chicago: Lake View Press, 1998.

Hirsch, Arnold R. *Making the Second Ghetto: Race and Housing in Chicago, 1940–1960.* Chicago: University of Chicago Press, 1998.

Hunt, D. Bradford. *Blueprint for Disaster: The Unraveling of Chicago Public Housing.* Chicago: University of Chicago Press, 2009.

Jacobs, Jane. *The Death and Life of Great American Cities.* New York: Random House, 1961.

Marciniak, Ed. *Reclaiming the Inner City: Chicago's Near North Revitalization Confronts Cabrini-Green.* Washington, D.C.: National Center for Urban Ethnic Affairs, 1986.

Vale, Lawrence J. *Purging the Poorest: Public Housing and the Design Politics of Twice-Cleared Communities.* Chicago: University of Chicago Press, 2013.

CHAPTER FIVE: THE MAYOR'S PIED-À-TERRE

Jane Byrne's stay at Cabrini-Green was a sensation, the subject of a thousand reports in the media that explored it from all angles. Every person I spoke with who lived or worked at Cabrini-Green or even followed the news during that time had a Mayor Byrne story or opinion to share. The Media Burn Archive has great documentary footage of Byrne's stay at Cabrini-Green. In addition to interviewing Dolores Wilson for this chapter, I also learned a good deal from conversations with Tara and Guana Stamps, Slim Coleman, Helen Shiller, Demetrius Cantrell, Jimmy Williams, Carol Steele, and Charles Price. And also from the sources listed below.

Byrne, Jane. *My Chicago.* New York: W. W. Norton, 1992.

Cohen, Adam, and Elizabeth Taylor. *American Pharaoh: Mayor Richard J. Daley: His Battle for Chicago and the Nation.* Boston: Little, Brown, 2000.

Hampton, Henry, et al. *Eyes on the Prize II: History of the Civil Rights Movement from 1965 to the Present.* Alexandria, VA: PBS Video and Blackside, Inc., 1990.

Marciniak, Ed. *Reclaiming the Inner City: Chicago's Near North Revitalization Confronts Cabrini-Green.* Washington, D.C.: National Center for Urban Ethnic Affairs, 1986.

Stamets, Bill. *Chicago Politics: A Theatre of Power.* Digital file. Chicago: 1987.

United States Commission on Civil Rights. Illinois Advisory Committee. *Housing,*

Chicago Style: A Consultation Sponsored by the Illinois Advisory Committee to the United States Commission on Civil Rights. Washington, D.C.: The Commission, 1982.

CHAPTER SIX: CABRINI-GREEN RAP

I based parts of this chapter on the long interviews I had with Annie Ricks, and on my interviews with Kelvin Cannon, Demetrius Cantrell, Jimmy Williams, Doug Shorts, Jesse White, Jackie Taylor, and other Cabrini-Green residents who had small parts in *Cooley High*. The tension on the set of *Good Times* was documented at the time in interviews given by cast members. Nicholas Lemann's book, cited below, helped as well, and was somewhat of an inspiration for my own book.

Lemann, Nicholas. *The Promised Land: The Great Black Migration and How It Changed America.* New York: Vintage Books, 1992.

CHAPTER SEVEN: CONCENTRATION EFFECTS

Kelvin Cannon described for me his arrest and time in prison, and parts of his biography were corroborated by other residents, police officers, and written accounts that celebrated his rise from unpromising origins. By the 1980s, Cabrini-Green was a fixture in the civic imagination, so I was able to use numerous reports on its further decline amid the revitalization of the surrounding central communities. The developers of Atrium Village toured me around the development. I learned more about the activism around the Chicago 21 plan and Harold Washington's ascent to the mayor's office from Helen Shiller (and from the *Keep Strong* leftist magazine she helped publish), Jesse Jackson, Jon DeVries, and Bill Stamets's amazing footage on Chicago politics which I cite below. I'd have to do a more careful survey of my notes to tally up all the city officials and consultants who cited William Julius Wilson's work on the deleterious effects of concentrated poverty as their reason to push for the demolition of Chicago's high-rise public housing.

Bennett, Larry. *The Third City: Chicago and American Urbanism*. Chicago: University of Chicago Press, 2010.

Grimshaw, William J. *Bitter Fruit: Black Politics and the Chicago Machine, 1931–1991*. Chicago: University of Chicago Press, 1992.

Hunt, D. Bradford. *Blueprint for Disaster: The Unraveling of Chicago Public Housing*. Chicago: University of Chicago Press, 2009.

Kleppner, Paul. *Chicago Divided: The Making of a Black Mayor*. DeKalb, Ill.: Northern Illinois University Press, 1985.

Marciniak, Ed. *Reclaiming the Inner City: Chicago's Near North Revitalization Confronts Cabrini-Green*. Washington, D.C.: National Center for Urban Ethnic Affairs, 1986.

Rivlin, Gary. *Fire on the Prairie: Chicago's Harold Washington and the Politics of Race*. New York: H. Holt, 1992.

Sampson, Robert J. *Great American City: Chicago and the Enduring Neighborhood Effect*. Chicago: University of Chicago Press, 2013.

Squires, Gregory D., et al. *Chicago: Race, Class, and the Response to Urban Decline*. Philadelphia: Temple University Press, 1987.

Stamets, Bill. *Chicago Politics: A Theatre of Power*. Digital file. Chicago: 1987.

Whitaker, David T. *Cabrini-Green in Words and Pictures*. Chicago: W3, 2000.

Wilson, William J. *The Truly Disadvantaged: The Inner City, the Underclass, and Public Policy*. Chicago: University of Chicago Press, 1987.

———. *When Work Disappears: The World of the New Urban Poor*. New York: Knopf, 1996.

CHAPTER EIGHT: THIS IS MY LIFE

I based parts of this chapter on my numerous conversations with Willie J. R. Fleming and other Cabrini-Green residents close to him. I interviewed Vince Lane, and I was able to learn more about his work as head of the CHA from discussions with reporters, researchers, and residents, and also from the extensive coverage of his controversial tactics and their meaning amid a perceived urban crisis. The following sources were also especially useful.

Burns, Ken, et al. *The Central Park Five*. DVD. [Arlington, Virginia]: PBS, 2013.

Didion, Joan. "New York: Sentimental Journeys." *The New York Review of Books*. January 17, 1991.

Hunt, D. Bradford. *Blueprint for Disaster: The Unraveling of Chicago Public Housing*. Chicago: University of Chicago Press, 2009.

Kotlowitz, Alex. *There Are No Children Here: The Story of Two Boys Growing Up in the Other America*. New York: Doubleday, 1991.

Popkin, Susan J., et al. *The Hidden War: Crime and the Tragedy of Public Housing in Chicago*. New Brunswick, NJ: Rutgers University Press, 2000.

Vale, Lawrence J. *Purging the Poorest: Public Housing and the Design Politics of Twice-Cleared Communities*. Chicago: University of Chicago Press, 2013.

CHAPTER NINE: FAITH BROUGHT US THIS FAR

The account of how residents of 1230 N. Burling came to manage their own high-rise is based on my conversations with Dolores Wilson and other tenants of the building; on records and reports archived at the CHA, the Metropolitan Planning Council, and the Burt Natarus papers at the University of Illinois at Chicago; on the documentary *Fired Up!*; and on the coverage that appeared in the media and on David Fleming's book listed below. For this chapter I also interviewed Rodnell Dennis, Eric Davis, James Martin, Peter Keller, and Veronica McIntosh.

Davis, Eric, et al. *The Slick Boys: A Ten-point Plan to Rescue Your Community By Three Chicago Cops Who Are Making It Happen*. New York: Simon & Schuster, 1998.

Fleming, David. *City of Rhetoric: Revitalizing the Public Sphere in Metropolitan America*. Albany: State University of New York Press, 2008.

Gangland, "Gangster City." History Channel, January 3, 2008.

Martin, James R. *Fired-Up!: Public Housing Is My Home*. Digital File. Oak Park, IL: Cineventure Inc., 1988.

CHAPTER TEN: HOW HORROR WORKS

This chapter is based in part on interviews with Annie Ricks and her family, Willie J. R. Fleming (along with his own personal video archive), Bernard Rose, Bill Tomes, Jim Fogarty, and many residents who worked with Brother Bill at Cabrini-Green.

Macek, Steve. *Urban Nightmares: The Media, the Right, and the Moral Panic Over the City.* Minneapolis: University of Minnesota Press, 2006.

Mann, Nicola. "The Death and Resurrection of Chicago's Public Housing in the American Visual Imagination." PhD dissertation: University of Rochester, 2011.

Martin, James. *My Life with the Saints.* Chicago: Loyola Press, 2006.

Richardson, Chris, and Hans Arthur Skott-Myhre. *Habitus of the Hood.* Chicago: Intellect, 2012.

Rose, Bernard, Virginia Madsen, Tony Todd, Xander Berkeley, Philip Glass, and Clive Barker. *Candyman.* DVD. Directed by Bernard Rose. [United States]: Columbia TriStar Home Entertainment, 2004.

CHAPTER ELEVEN: DANTRELL DAVIS WAY

The killing of Dantrell Davis was the biggest news story in Chicago in 1992, and the media frenzy is both a source for and topic of this chapter. I also interviewed Dantrell's mother, Annette Freeman, and numerous residents, reporters, and city officials who lived through the immediate aftermath. I learned more about the ensuing gang truce from Wallace "Gator" Bradley, Hal Baskin, Maurice Perkins, Prince Asiel Ben Israel, Tara and Guana Stamps, Kelvin Cannon, Eric Davis, James Martin, Patricia Hill, and Frederick "Hoggie Wolf" Watkins.

Bennett, Larry. *The Third City: Chicago and American Urbanism.* Chicago: University of Chicago Press, 2010.

Bulkeley, Kelly, et al. *Among All These Dreamers: Essays on Dreaming and Modern Society.* Albany, NY: State University of New York Press, 1996.

Cohen, Adam, and Elizabeth Taylor. *American Pharaoh: Mayor Richard J. Daley: His Battle for Chicago and the Nation.* Boston: Little, Brown, 2000.

Coyle, Daniel. *Hardball: A Season in the Projects.* New York: G.P. Putnam's Sons, 1993.

Michaeli, Ethan. *The Defender: How the Legendary Black Newspaper Changed America: From the Age of the Pullman Porters to the Age of Obama.* Boston: Houghton Mifflin Harcourt, 2016.

Obama, Barack. *Dreams From My Father: A Story of Race and Inheritance.* New York: Random House, 1995.

Pollack, Neal. "The Gang that Could Go Straight." *Chicago Reader*. January 26, 1995.

Shafton, Anthony. *Dream-Singers: The African American Way With Dreams*. New York: J. Wiley & Sons, 2002.

CHAPTER TWELVE: CABRINI MUSTARD AND TURNIP GREENS

I learned more about the 1230 N. Burling resident management corporation and the renovation of the high-rise from documents held in the CHA archives and from interviews with Dolores Wilson and other tenants in her building. Peter Benkendorf sent me the nearly complete run of *Voices of Cabrini*, and I spoke about the community newspaper with him, Mark Pratt, Pete Keller, and Jimmy Williams. The account of Cabrini-Green's fitful redevelopment was well-documented in the local media, and for my purposes the investigative work of the *Chicago Reporter* proved especially helpful. I also benefited from conversations with William Wilen, Marilyn Katz, Vince Lane, Carol Steele, and Richard Wheelock, and from the research on the different development plans shared with me by Larry Bennett. It's useful here to highlight Lawrence Vale's *Purging the Poorest*, a book I cite below and elsewhere. I benefitted greatly from Vale's account of the Cabrini-Green area as a "twice-cleared community" and particularly from his writing about its second erasure beginning in the 1990s.

Bennett, Larry, and Adolph Reed Jr. "The New Face of Urban Renewal: The Near North Redevelopment Initiative and the Cabrini-Green Neighborhood," in *Without Justice for All: The New Liberalism and Our Retreat From Racial Equality*, ed. Adolph Reed Jr. Boulder, Colo.: Westview Press, 1999.

———, Janet L. Smith, and Patricia A. Wright. *Where Are Poor People to Live?: Transforming Public Housing Communities*. New York: Routledge, 2006.

Fleming, David. *City of Rhetoric: Revitalizing the Public Sphere in Metropolitan America*. Albany: State University of New York Press, 2008.

Keller, Pete "Esaun." *Cross the Bridge*. Chicago: Self-published, 2012.

Pattillo, Mary E. *Black on the Block: The Politics of Race and Class in the City*. Chicago: University of Chicago Press, 2007.

Vale, Lawrence J. *Purging the Poorest: Public Housing and the Design Politics of Twice-Cleared Communities*. Chicago: University of Chicago Press, 2013.

Wilen, William P. "The Horner Model: Successfully Redeveloping Public Housing." *Northwestern Journal of Law & Social Policy* 62 (2006).

CHAPTER THIRTEEN: IF NOT HERE . . . WHERE?

This chapter is based in part on my interviews with Kelvin Cannon, Willie J. R. Fleming, Annie Ricks, and the family members and friends who knew them. I learned about specific details of redevelopment meetings held at Cabrini-Green from Matthew McGuire's dissertation cited below and from the documentary *Voices of Cabrini*. Several people added to my understanding of the Coalition to Protect Public Housing, including Carol Steele, Janet Smith, Regina McGraw, Bruce Orenstein, Jim Field, and Will Small.

Bennet, Larry, Janet L. Smith, and Patricia A. Wright. *Where Are Poor People to Live? Transforming Public Housing Communities*. New York: Routledge, 2006.

Bezalel, Ronit, and Antonio Ferrera. *Voices of Cabrini*. Digital File. Directed by Ronit Bezalel. Chicago, Ill.: Facets Video, 1999.

Ehrenhalt, Alan. *The Great Inversion: And the Future of the American City*. New York: Knopf, 2012.

McGuire, Matthew. "Chicago Private Parts: The Relationship Between Government, Community and Violence in the Redevelopment of a Public Housing Complex in the United States." PhD dissertation: Harvard University, 1999.

Royko, Mike. *Boss: Richard J. Daley of Chicago*. New York: Dutton, 1971.

Vale, Lawrence J. *Purging the Poorest: Public Housing and the Design Politics of Twice-Cleared Communities*. Chicago: University of Chicago Press, 2013.

CHAPTER FOURTEEN: TRANSFORMATIONS

In addition to relying on the sources listed below and on considerable media coverage, my reporting on the Plan for Transformation included conversations with Richard M. Daley, Julia Stasch, Joseph Shuldiner, Sudhir Venkatesh, William Wilen, Alex Polikoff, Carol Steele, Richard Wheelock, Walter Burnett, Robert Whitfield, Lewis

Jordan, Marilyn Katz, and many others. Thomas Sullivan's reports on the shortcomings of the Plan for Transformation were also useful. For the section of this chapter on Willie J. R. Fleming, I interviewed him and relied as well on court and police records.

Bennet, Larry, Janet L. Smith, and Patricia A. Wright. *Where Are Poor People to Live?: Transforming Public Housing Communities*. New York: Routledge, 2006.

Bezalel, Ronit, Catherine Crouch, Judy Hoffman, Brenda Schumacher, Marguerite Mariama, Janet L. Smith, Deidre Brewster, Mark Pratt, D. Bradford Hunt, and Duane Buford. *70 Acres in Chicago: Cabrini Green*. DVD. Directed by Ronit Bezalel. 2015.

Fennell, Catherine. *Last Project Standing: Civics and Sympathy in Post-Welfare Chicago*. Minneapolis: University of Minnesota Press, 2015.

Fleming, David. *City of Rhetoric: Revitalizing the Public Sphere in Metropolitan America*. Albany: State University of New York Press, 2008.

Hunt, D. Bradford. *Blueprint for Disaster: The Unraveling of Chicago Public Housing*. Chicago: University of Chicago Press, 2009.

Kalven, Jamie. "Kicking the Pigeon." *The View from the Ground*. 2005, 2006.

Venkatesh, Sudhir Alladi. *American Project: The Rise and Fall of a Modern Ghetto*. Cambridge, Mass: Harvard University Press, 2000.

———. *Chicago Public Housing Transformation: A Research Report*. New York: Center for Urban Research and Policy, Columbia University, 2004.

———, and Larry Kamerman. *Dislocation*. DVD. [S.l.]: Alladi Group, 2005.

CHAPTER FIFTEEN: OLD TOWN, NEW TOWN

In addition to interviews with Kelvin Cannon, Dolores Wilson, and Annie Ricks, this chapter is based on documents and letters provided to me by Cannon, court records of the contested tenant council election at Cabrini-Green, and the recollections of numerous people who lived and worked at Cabrini-Green and Parkside of Old Town, including Carol Steele, Charles Price, Peter Holsten, Abu Ansari, Deirdre Brewster, Richard Sciortino, Kenneth Hammond, and Tyrone Randolph. Starting in 2010, I also started attending public meetings at Cabrini-Green and other forums held by the CHA at which the dynamic in the new mixed-income developments was always a point of contention. Ronit Bezalel's documentary *70 Acres in Chicago* was also very useful, as was talking with her as we both worked on this subject.

Bennet, Larry, Janet L. Smith, and Patricia A. Wright. *Where Are Poor People to Live?: Transforming Public Housing Communities.* New York: Routledge, 2006.

Bezalel, Ronit, Catherine Crouch, Judy Hoffman, Brenda Schumacher, Marguerite Mariama, Janet L. Smith, Deidre Brewster, Mark Pratt, D. Bradford Hunt, and Duane Buford. *70 Acres in Chicago: Cabrini Green.* DVD. Directed by Ronit Bezalel. 2015.

———, and Antonio Ferrera. *Voices of Cabrini.* Digital File. Directed by Ronit Bezalel. Chicago, Ill.: Facets Video, 1999.

Chaskin, Robert J., and Mark L. Joseph. *Integrating the Inner City: The Promise and Perils of Mixed-income Public Housing Transformation.* Chicago: University of Chicago Press, 2015.

Vale, Lawrence J. *Purging the Poorest: Public Housing and the Design Politics of Twice-Cleared Communities.* Chicago: University of Chicago Press, 2013.

CHAPTER SIXTEEN: THEY CAME FROM THE PROJECTS

I attended the vigil outside 1230 N. Burling the night before the start of its demolition. Jan Tichy provided me with the poetry included in his Project Cabrini Green and with photographs he took inside the cleared units of the high-rise. Dolores Wilson and Annie Ricks recounted their moves and showed me the public housing developments where they had been relocated. There was extensive reporting on the relocations of public housing families and the increase in violence in the neighborhoods where they ended up. I also interviewed Lewis Jordan, Eric Davis, and others about these moves.

Feldman, Roberta M., Sheila Radford-Hill, and Susan Stall. *The Dignity of Resistance: Women Residents' Activism in Chicago Public Housing.* New York: Cambridge University Press, 2004.

Rosin, Hanna. "American Murder Mystery," *The Atlantic*, July/August 2008.

CHAPTER SEVENTEEN: THE PEOPLE'S PUBLIC HOUSING AUTHORITY

I was present for many of the events I describe in this chapter, and I benefited as well from interviews with Willie J. R. Fleming, Toussaint Losier, Shirley Henderson, Martha Biggs, Thomas Turner, Edward Voci, Rahm Emanuel, Carol Steele, Patricia Hill, and Emma Harris.

Emanuel, Ezekiel J. *Brothers Emanuel: A Memoir of an American Family.* New York: Random House, 2013.

Gottesdiener, Laura. *A Dream Foreclosed: Black America and the Fight for a Place to Call Home.* Westfield, N.J.: Zuccotti Park Press, 2013.

CHAPTER EIGHTEEN: THE CHICAGO NEIGHBORHOOD OF THE FUTURE

I witnessed firsthand many of the changes to the Cabrini-Green neighborhood in the past decade. (I happened to attend the East Bank Club luncheon titled "From Cabrini-Green to NoCa.") Since 2012 I've been going intermittently to the Near North Unity Program monthly meetings, and I was present for a number of the public forums to discuss the merger between Cabrini-Green's Jenner Elementary and the Gold Coast's Ogden International School. I have visited both schools as well and spoken to parents and teachers from both communities. Jesse White let me sit in as he met with constituents during a weekly ward committeeman night, an evening when Kelvin Cannon was dutifully present. I also spent time with Annie Ricks and her family as they tried to move out of Wentworth Gardens. I visited Ricks in the hospital; I was there when she passed away, and I attended her funeral. Likewise, I was on the boat ride with J. R. Fleming, Raymond Richard, and Brother Jim that I describe in the book's final pages. Although I'm born and raised in Chicago, and I live on the city's South Side, it was my first time on the river.

Index

About the Author

BEN AUSTEN has written for many publications, including *Harper's Magazine*, the *New York Times Magazine*, *GQ*, and *New York* magazine. He lives in Chicago.